The bowels of the earth

JOHN ELDER

The bowels of the earth

OXFORD UNIVERSITY PRESS

1976

Oxford University Press, Ely House, London W.1

GLASGOW NEW YORK TORONTO MELBOURNE WELLINGTON
CAPE TOWN IBADAN NAIROBI DAR ES SALAAM LUSAKA ADDIS ABABA
DELHI BOMBAY CALCUTTA MADRAS KARACHI LAHORE DACCA
KUALA LUMPUR SINGAPORE HONG KONG TOKYO

CASEBOUND ISBN 0 19 854412 X

© OXFORD UNIVERSITY PRESS 1976

PRINTED IN GREAT BRITAIN BY
THOMSON LITHO LTD., EAST KILBRIDE SCOTLAND

Preface

WRITING a book such as this is like arranging a tour of a very large art gallery. Some people want to dash past every picture in the place, others prefer just to look at a few masterpieces. I have chosen neither of these approaches. My tour takes one strong aspect of the subject—the role of scale—and follows it from large to small systems. As a result some rooms are not seen at all, and even some great masterpieces and contemporary disasters are ignored.

The material is presented in 15 chapters. Chapter 1 is an introduction. Chapters 2–6 are about the earth as it is, in the broadest terms—global geology. Chapters 7–12 deal with macro-dynamics—the behaviour of systems generally of greater scale than the thickness of the upper mantle. Chapters 13–15 are a selection of topics on meso-dynamics—more or less self-contained systems resident in the upper mantle and crust.

The first few chapters deal with matters on the largest scale and ask only the slightest of questions. In later chapters we see the highly self-interactive nature of the earth and realize that many of our early ideas were rather naïve. I prefer this gradual unfolding of the story rather than a treatment in which every part is handled on the same level. Further, I have taken considerable liberty with ordering the material, which is presented as if it were discovered in the order given, whereas often it didn't happen like that at all.

Four features of the arrangement of the book should be noted.

1. There is no explicit reference in the text to the diagrams and tables. Rather, these are exhibits, placed near the relevant text.

2. There is an extensive glossary–index, a kind of micro-encyclopedia. This avoids the distraction of having the text cluttered with important but minor items of information. The glossary includes enough details for this book as well as some information that should be helpful to the beginner when reading other related books.

3. More elaborate discussions of particular models are labelled 'Theoretical sketch'. This is an attempt to cope with the problem of variation in readers' backgrounds.

4. There are no references given in the body of the text. This was not an easy decision. The literature of the subject is vast and usually many people have contributed to each part of it. The proportion that could be

referenced in a small book like this one would give quite a false impression. The source material used in this book is taken from the books and articles referred to in the reading list and glossary–index.

Who is this book written for other than myself?

The man in the street.　You should see one aspect of modern geology very clearly: our understanding of dynamical processes inside the earth—a sort of meteorological or physiological insight. The glossary–index will be especially helpful, not only for this book but for others which cover related topics. I hope you won't be put off by rather a lot of mathematical formulae.

The student.　Perhaps you are studying science at school or are a university undergraduate. You will find the material in this book useful as a supplementary text in understanding the broader dynamical aspects of the earth's interior. All the topics are treated with a minimum of description but are all quite quantitative. The problems (p. 185), many of which are very difficult, are for you to tackle. I hope you find them stimulating.

The teacher.　This is not a course textbook, but I expect you will find useful teaching material in it. Any teacher of a natural science faces an awful dilemma. The great scope of the subject and the technical and professional requirements demand an enormous amount of purely descriptive material. Unfortunately this tends to leave too little time for experimentation and the strictly scientific as opposed to the technical side of the subject. Teachers recognize this difficulty all too well, but should not be content with just 'tarting up' the subject with descriptions of the latest fashionable ideas. Let us take a lesson from our maths, physics, and chemistry colleagues and leave most of the description in books, making our students spend their time doing experiments, simulations, and problems. I don't pretend that this book is the answer to these difficulties, but I hope that its approach will suggest useable teaching devices to you.

The expert.　I have very little here for you. None of the material is new, but some of it is in a new guise. Of this, perhaps the most novel is the thermal history story of Chapter 9 (a nice elaboration of some work of mine done in 1967) and the ideas on magmatic discharge rates in Chapter 14 (which is simply a 'translation' of my work, done in 1960 and published in 1966, on fumaroles and geysers). Although an elementary book, this is an attempt to give a quantitative and consistent scenario of the range of interior physical processes. I hope you won't mind too much when you see me skating on thin ice.

This book is based on material that I have used since 1970 in short courses on geophysics and macrogeology which form a part of the undergraduate geology programme given by members of the Department of Geology at

Manchester University. During the lectures I have given on this material I have, both for fun and with serious purpose, passed round the class various objects which are referred to as 'the earth'. For example, on the first day, after pretending to walk through the solar system, I pick up 'the earth', and say 'This is the earth. Mm, it is quite massive, and rather squashy. Here, you feel it, but be careful not to damage it.' The object is a bladder filled with sand. For the lecture on the earth as a jumping bean I use a large flat aluminium annulus. The hot earth calls for a brick saturated with hot water. And so on. The idea is to get used to thinking about the earth just as you do about any other object, to feel that it is as familiar to you as a stone held in your hand. Many objects around you will suggest analogues with the earth or one of its parts; the kitchen abounds with opportunities. As you go through this book try to think of something in your everyday experience which has properties a bit like those of that particular aspect of the earth. It is often much easier to think about a simple analogue. Of course, a multi-layered cake is not the same as the earth, but there may be things that you notice in a cake that suggest useful ideas about the earth. Even so, in this book, I am more interested in the recipe than the cake.

J. W. E.

Department of Geology
Manchester University
November 1974

Acknowledgements

I HAVE DRAWN on a wide range of material from textbooks and journals. Most of this material is readily available in the items of the reading list and glossary–index. I am grateful to the following authors and publishers for supplying photographs or giving permission to use their visual material (any minor alteration or adaptation is my own): Fig. 4.1: Geodetic Institute, Copenhagen; Figs. 8.5, 8.7, 8.8(b), 8.9(a), (b): J. W. Deardorff; Fig. 12.2: C. J. Cambell; Fig. 12.3: J. Fitch; Figs. 12.5, 14.6: H. Ramberg; Fig. 13.5: N.Z.D.S.I.R.; Fig. 13.9: the Royal Society London; Fig. 13.12: B. I. Nielsen; Fig. 14.2: P. G. Harris; Fig. 14.11: Fred Bullard (University of Texas Press); Fig. 15.1: T. Huntingdon.

Two areas of my own experience, work in the field and teaching undergraduate geology students at Manchester, have dominated the writing of this book and I wish to express my gratitude to the many people involved, especially to my friends and colleagues in the Department of Geology at Manchester University, the New Zealand D.S.I.R., the Geological Survey of Japan, the Societa Lardarello, and the Greenland Geological Survey. By no means least, many individuals have helped with the task of converting the original manuscript into its present form. I will mention only a few: Patricia M. Crook who typed it, Sue Maher who made some lovely photographs, and Dr. Bill Sowerbutts who read and commented on it. I should also like to thank the staff of the Oxford University Press for their help and encouragement during the writing of this book.

Notes on reading this book

1. Technical terms which are not defined in the body of the text are defined in the glossary–index. Symbols are defined there too. The glossary should be frequently used by the layman.

2. Information that is common knowledge and can be found in a standard dictionary or small encyclopedia is not defined in the text or the glossary. I have in general used the rule that where an item of information can be checked or followed up by reference to the *Encyclopaedia Brittanica* (1974), I have given explicit information only if the item is important to the argument.

3. Where numerical values are needed for the magnitude of quantities, standard values as given in the glossary are used, unless stated to the contrary, and are not repeated explicitly in the text. Many of these values are known to adequate accuracy but others are either unknown or, at best, known only to an order of magnitude. This is noted together with the often somewhat arbitrarily chosen value to be used here.

4. Most quantities are quoted in System International (SI) units using: m, metre; kg, kilogram; s, second; K, degree Kelvin and multiples of these units. Electrical and optical quantities are not needed in this book. SI additional units and their multiples are used for:

angle, time, mass (ton $= 10^3$ kg), pressure (bar $= 10^5$ N m^{-2}), temperature (°C), viscosity (P, poise $= 0.1$ kg m^{-1} s^{-1}).

I also use d, 1 day $= 86\,400$ s and yr, 1 year $= 3.156 \times 10^7$ s.

Densities are always given in the units 10^3 kg m^{-3} $-$ g cm^{-3} $=$ ton m^{-3} and never in multiples.

5. Some help in coping with the variety of units used in existing literature is given in the glossary. In geophysics the commonest other units are those of the centimetre–gram–second–calorie system.

Prelude: A glimpse of neolithic cosmogony

An extract from Chapter IV, *The Maori*, by ELSDON BEST, 1924 (2 vols.) (H. H. Tombs, Wellington).

IN MIST-LADEN DAYS of the remote past the sky and earth were not parted as we now see them, for Rangi the Sky Father closely embraced Papa the Earth Mother. All was darkness between them, no light existed, nothing could mature, nothing could bear fruit, all things merely existed, or moved aimlessly about in a realm of darkness. When the children [the forces of nature] of these primal parents were born they found themselves dwelling in darkness, and clung to the body of the Earth Mother, sheltering within her armpits. [The Unknowable] prevailed.

The offspring soon became discontented with their lot in the world. The conditions of life were irksome and unpleasing, so cramped they were for space. This lack of space was the result of the close contact of their parents at that remote period, for Rangi still embraced Papa; sky and earth were close together. It was Tane [light] who proposed to separate them, saying: 'Let us part our parents; let us force Rangi upward, suspend him on high, and let Papa lie in space'. That task proved to be a difficult one, so closely did the parents cling together in their great affection for each other. It was found to be necessary to sever the arms of Rangi ere he could be forced upward. The blood from his grievous wounds flowed over, and was absorbed by the body of Papa, hence the horu or red ochre found within her body even unto this day.

A time came when the grief of Papa on account of her separation from Rangi, her old-time love, came to be known to Io in the uppermost heaven. The sound of her wailing was borne upward, hence Io sent Ruatau down to seek the cause of the ceaseless lamentation. Io now commanded that the Earth Mother be turned over, so that she might no longer gaze upon her lost love Rangi. This is known as the Hurihanga a Mataaho, the overturning by Mataaho.

Even so was the Earth Mother turned over, so that she lay face down to Rarohenga, the underworld, hence man now dwells on her back instead of on her breast, as of yore. When she was so turned over, her youngest child, Ruaumoko, was still at her breast.

This child she was allowed to retain in her solitude. The brothers of Ruaumoko resolved to grant him some comfort in his dark realm, hence they gave the boon of fire. This fire was obtained for the purpose from Raka-hore, the personified form of rock. This subterranean fire is known as ahi komau, buried fire. Ruaumoko is responsible for all volcanic outbursts and earthquakes. [The first syllable of his name is the common term for an earthquake; it means 'to shake'.]

Tane despatched Tawhirimatea [wind] to procure the Cloud Children, who sprang from the warmth and perspiration of the body of the Earth Mother. And so the Wind Children were sent to fetch them. They brought Ao-nui and Ao-roa [Great Cloud, Long Cloud] and all the numerous Cloud Children to serve as a garment to cover the body of the Sky Parent. Such are the clouds above us. The body of the Earth Mother was also covered, and the garment bestowed upon her was composed of vegetation, which protected and warmed her.

Contents

1. Global jam-pot

OUR EARTH is a vast machine, insatiably reconsuming itself.

But when we first approach the earth from space it appears merely as a small stone. And as you would look at a stone you have just picked up, so do we examine the earth. For example, we can measure it and find out how much matter there is in it. But we can't break it open! And since we have only one such stone and it is too big to fiddle about with very much, we must infer what is inside it. But how are we to do that? Whereas most laboratory-scale sciences have the opportunity of isolating a system of interest and repeatedly stimulating it, we are obliged (most of the time through sheer necessity) to work with an analogous system—a model. With such models we can try to simulate the patterns found in nature. Our model becomes a theoretical tool, and its validity is measured by comparing it quantitatively with our observations of the natural system. Each model can reproduce only a small number of aspects of the actual system—but that is just what we want. Here we are simply playing the game of the laboratory scientist by isolating a process for detailed analysis. In fact there really are no more than superficial differences between these methods: we can no more go directly into an atom or gene than we can go to the centre of the earth.

Since the earliest times man has puzzled about himself and his surroundings. In recent years we have seen our planet for the first time from the outside. As yet we have not seen our planet from the inside. Our knowledge of the interior of the earth and the processes inside it which form the physiognomy of its surface is slight. Nevertheless, this book is an attempt to show you a view from the inside, a view through one pair of eyes.

But what we see depends on what we look at. This is especially true for large complex objects. The scale of a system can have a powerful effect on its behaviour. If, for example, we take some clay and bake it, we can make a brick. Chemically the brick is difficult to distinguish from the clay, but it is nevertheless quite a different thing. Now take some bricks and make a building. This is still really clay but that is almost irrelevant. Again let us take some water, air, heat, and electricity and make a cumulus cloud. Somehow these things have acquired characteristics which are more than the sum of their parts. How is this possible? Clearly the fabricated system has more degrees of freedom—the opportunity to manifest a greater variety of organization. Provided the parts or ingredients of our fabricated system are able to interact with each other, then not only do more complex patterns develop but quite new patterns of behaviour arise. Clearly, even if we knew

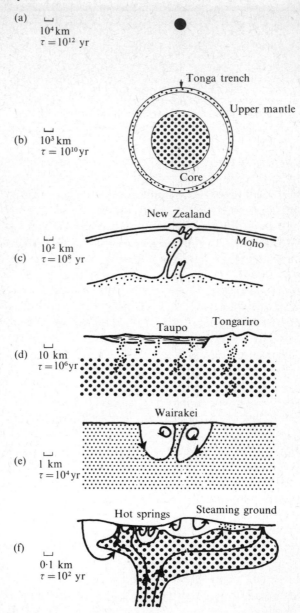

(a)   ⊔
10⁴ km
$\tau = 10^{12}$ yr

(b)   ⊔
10³ km
$\tau = 10^{10}$ yr

(c)   ⊔
10² km
$\tau = 10^8$ yr

(d)   ⊔
10 km
$\tau = 10^6$ yr

(e)   ⊔
1 km
$\tau = 10^4$ yr

(f)   ⊔
0·1 km
$\tau = 10^2$ yr

Fig. 1.1. Diagrammatic sections of the earth. The length scales of successive diagrams are in the ratio 1:10. (a) featureless stone; (b) the whole earth, showing mantle and core; (c) the upper mantle, near New Zealand; (d) the crust of the Taupo area; (e) the upper crust and the Wairakei hydrothermal system; (f) the surface zone of the crust. The time-scales are very roughly in the ratio 1:100.

more or less all there was to know about water, air, heat, and electricity as such, still we wouldn't know how a cumulus cloud works. Thus we recognize that there are aspects of the macro-environment which require descriptions quite different from those appropriate to the ingredients. This notion is often inadequately summed up by saying, 'We can't see the wood for the trees'.

Look at the earth through the eye of the geologist. Contract your time and distance scales so that 10^{10} years becomes an hour and the earth has shrunk until its diameter is 1 m. You will see the surface of the earth in a vigorous state of motion, rather like a large pot of jam cooking on a stove, with bits of scum moving erratically around in the surface, blobs popping up to the surface, and the body of the fluid moving up and down in irregular eddying motions. Let me show you the jam-pot of the earth.

Heat is transmitted through the deep layer of jam by the large-scale eddying motions. Moving about in the surface of our jam are a few patches of scum. These are our continents. Sometimes these patches are torn apart, later to be rejoined to others. Occasionally we see blobs of less viscous jam break through and ooze over the surface of the more viscous jam. This process is most clearly seen shortly after we place the pot on the stove. The regions of surface mass discharge are volcanic systems. Heat leaves our jam-pot by warming the ambient air and by losing water vapour. If we look closely at the surface we occasionally see a strong puff of vapour where this process is greatly intensified. These regions of discharge of water and water vapour in steaming ground, hot springs, geysers, fumaroles, mud volcanoes, and phreatic explosions are called thermal areas.

In order to describe these processes, we introduce, at each scale, a quantity which dominates the behaviour of structures of that scale. These quantities effectively identify the nature of the working material. Only six such quantities will be used in this book: size, mass, density, viscosity, temperature, and permeability. Thus the strategy of this book is to consider a sequence of scales with a progressively more elaborate specification of the working material. At each stage we must adjust our geological eye to the length- and time-scale of the phenomenon of interest. Let us begin with the largest scale, by setting our eye to 10^4 km and 10^{12} years.

FIG. 2.1. The earth from space.

2. Stone in space

Where is it?

WHERE is the earth? This would be a difficult question to answer for a hypothetical space-traveller from another part of the cosmos. Notable objects within 2×10^{19} km of the earth are the great spiral galaxy in Andromeda and (quite near) the two Magellanic clouds. Towards the outer rim of our galaxy there is a yellow G-type star surrounded by nine planets. The third one, at a distance of $1\cdot5 \times 10^{8}$ km from its sun, is the earth: a planet of typical size, of mass 6×10^{24} kg, with a solitary satellite.

How old is it?

The visible universe as revealed by astronomical observation of stars and galaxies has been interpreted as a system that has been steadily developing for a period in excess of 10^{10} years. Our own sun coalesced and started its nuclear furnace 5×10^{9} years ago.

The oldest rocks so far found on earth are granodioritic gneisses from West Greenland; their age is about 4×10^{9} years. Since these must themselves have been derived as sediments from earlier rocks, the granitic crust of the earth has existed for longer than this. In this book a nominal age of 5×10^{9} years is used. This point in time is inevitably a little vague since it refers to a time before which processes on an astronomical scale were dominant in the formation of the earth; after it, geological processes within a distinct body were dominant. The instant in time when a distinct surface first formed, across which was a clear density discontinuity, is the beginning of geological time.

Shape and size: geodesy

The earth is nearly spherical; the sphere of volume equal to that of the earth has a radius of 6371 km. Careful measurement gives the over-all shape as closer to an ellipsoid of revolution about the axis of daily rotation, of ellipticity 1/298. The equatorial and polar axes are 6378·2 km and 6356·8 km respectively, with large-scale departures from this shape of typically one part in 10^{5}. The interior of the earth is thus close to being in hydrostatic equilibrium.

Direct observation from a space-vehicle unquestionably shows that the earth is nearly spherical. This has been more-or-less known since ancient times from various observations; for example, the earth's shadow during a lunar eclipse has the shape of a circular arc.

Quantitative measurement uses the simple device of finding the length of arc l between two places A and B on the earth's surface and the angle θ between plumb-lines at A and B. The local radius of curvature r is then given by l/θ, where θ is measured in radians (1 rad = 57·3°). The arc can be obtained in a variety of ways: Eratosthenes (276–192 B.C.) used the estimated distance travelled in a day by a camel train; Poseidonius

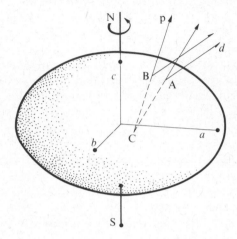

FIG. 2.2. Sketch of ellipsoid of flattening $\frac{1}{3}$ obtained by squashing a sphere so that the polar axis is $\frac{2}{3}$ the equatorial axis. The earth is squashed only 1 per cent of this. C = centre of curvature of arc AB; p, plumb-line; d, to distant object. Arc length l = AB, angle θ = ACB in radians, radius of curvature r = AC = BC related by $l = r\theta$.

(135–50 B.C.) an estimate of the speed of a sailing ship. In modern times, following Snell and Picard, we use the method of triangulation employed by surveyors. The angle is most simply obtained if A and B lie north and south of each other by observing a celestial object when it is north or south of the observer—for example, the sun at noon in midsummer. Modern instruments, including very accurate clocks, allow these measurements to be made to somewhat better than one part in 10^6; we can determine r to within a few metres.

Theoretical studies of Newton and Huygens suggested that a homogeneous self-gravitating mass would be spherical; but if it were rotating uniformly, centrifugal forces would bulge the body into an oblate spheroid. The first unequivocal measurements of the departure from sphericity were made in 1735–6 by comparing arcs in Ecuador, France, and Lapland. A more precise determination during 1792–8 followed the definition then of a new unit of length, the metre, defined as 10^{-7} of the distance from the pole to the equator on the meridian through Paris. We now know that this work was about 0·02 per cent out; the meridional quadrant is 10 002 km. Nevertheless this measurement showed that the earth was squashed along the polar axis. Modern data give the 'flattening' $e = (a-c)/a \approx 1/298{\cdot}25$. Thus, $1°$ of latitude corresponds to $110{\cdot}6$ km at the equator and to $111{\cdot}7$ km at the poles. The difference between these two figures, $1{\cdot}1$ km, is quite large!

How much matter?

On the human scale we readily obtain the mass of an object by comparison with other masses, either directly by using a balance or indirectly by using a calibrated spring. To estimate the mass of the earth two sets of measurements are necessary: a calibration measurement in the laboratory and measurements on the earth and its satellite.

The force of gravity acting between two masses m, M a distance r apart was found by Newton to be GmM/r^2, where G is a universal constant. In the laboratory, the force between two massive balls, first measured directly by Cavendish in 1791, gives $G = 6{\cdot}67 \times 10^{-11}\,\mathrm{N\,m^2\,kg^{-2}}$. This single entity characterizes our universe.

If m is the mass of a satellite orbiting a much more massive partner in a nearly circular orbit of radius r, the centrifugal force mv^2/r, where v is the orbital velocity, must balance the gravitational force. If T is the time for one revolution of the satellite around its orbit, noting that $vT = 2\pi r$ we obtain an expression for the mass of the central body: $M = 4\pi^2 r^3/GT^2$, a relation independent of the mass of the satellite. For this we thank Messrs Kepler and Newton.

For the earth and its satellite the moon $T = 27{\cdot}322$ d (d = days), $r = 3{\cdot}84 \times 10^5$ km, and hence $M = 6 \times 10^{24}$ kg. In a similar manner by considering the earth as a satellite of the sun, the mass of the sun is found to be 2×10^{30} kg. The mass of the earth is minute, only 3×10^{-6} that of the sun.

Moments of inertia: the spinning top

If the earth were a homogeneous sphere its moment of inertia about any axis would be $0{\cdot}4\,Ma^2$, where M is the mass and a is the radius of the sphere. For a non-uniform body let A, B, C be the principal moments of inertia. For the earth, if C is the moment about the polar axis, the moments A and B about equatorial axes are nearly equal, that is, $|A-B|/C \ll 1$. Estimates of A and C can be obtained in several ways. For example, the shape of the earth determines a value of $(C-A)$. Modern determinations use astronomical observations. The earth behaves like a spinning top and its axis of rotation precesses about a cone of half-angle $23°27'$ in $25\,735$ years. Hence, $(C-A)/A = 3{\cdot}275 \times 10^{-3}$. Incidentally this shows that C and A are nearly equal—which is not surprising when the earth is so nearly spherical. Similarly the orbital plane of a satellite precesses. For example, in a typical case of an artificial satellite at a height of 600 km the normal to the orbit might precess round a cone of half-angle 20° at a rate of 6·5° per day. This rate can be measured to somewhat better than 0·1 per cent, giving $(C-A)/Ma^2 = 1·0827 \times 10^{-3}$. Combining these two results, we obtain $C = 0·3309\,Ma^2$.

It is important not to be fooled by the equatorial fattening of the earth. This arises, as far as present measurements can determine, solely from the centrifugal bulging of a hydrostatic rotating earth. The residual moments of inertia after this contribution is removed are $(C-A)/Ma^2 \approx 2 \times 10^{-5}$ and $(C-B)/Ma^2 \approx 1 \times 10^{-5}$. Thus, if the earth's rotation were turned off and the equatorial bulge allowed to subside, but nothing else, the object would be rather like a rugby football: C about the long axis and A about an axis through the laces. For the earth, the laces lie in the meridional plane 165° E–015° W.

The interior density

The measurements of size, mass, shape, and moment of inertia allow us to make our first deduction about the interior.

Let us first think of an earth of uniform density and rotating at uniform angular velocity. Maclaurin showed that a homogeneous non-rigid globe of

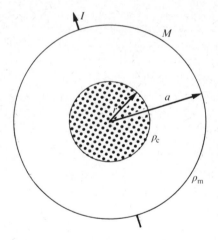

FIG. 2.3. Egg-model diagram: section through a spherical two-layer earth.

the same size, mass, and angular velocity as the earth would have flattening $e \sim 1/230$. This is too large: the actual value corresponds to a larger density, which is possible only if there is a substantial increase of density towards the centre of the earth. Modern calculations which allow for radial density variation give estimates of e to within 0·3 per cent.

Similarly, since the moment of inertia C is smaller than the corresponding value for a homogeneous body, there must be a mass concentration towards the centre of the earth.

Finally, given the mass $M = 5{\cdot}975 \times 10^{24}\,\text{kg}$ and the volumetric radius $a = 6371\,\text{km}$, the mean density $\rho = M/\tfrac{4}{3}\pi a^3 = 5{\cdot}517\,\text{g cm}^{-3}$. Now most rocks from the earth's surface have densities in the range $2{\cdot}7$–$3{\cdot}3\,\text{g cm}^{-3}$. Hence there must be a substantial increase in density inwards.

Theoretical sketch: egg model

Let us now put this information together into a simple spherical model.

As a first attempt at representing this density change consider an egg-like earth: an outer region of density ρ_m, the white, or mantle; an inner region of density ρ_c, the yolk, or core. In each region we assume that the density is uniform. We shall use our knowledge of the mass of the earth, by using the mean density ρ, and its moment of inertia C to find possible relationships between ρ_m, ρ_c, and the radius r of the core. Calculating the mass and moment of inertia separately for the two regions we have, for any given r, two equations which are readily solved for ρ_m, ρ_c:

$$\rho = \rho_m + (\rho_c - \rho_m)(r/a)^3,$$

$$\xi\rho = \rho_m + (\rho_c - \rho_m)(r/a)^5,$$

where
$$\xi = 0{\cdot}3309 \times \tfrac{5}{2} \text{ from the value of } C.$$

For values of $r < 2000\,\text{km}$, ρ_m is nearly constant but ρ_c is rising to very large values. For values of $r > 5000\,\text{km}$, ρ_c is nearly constant but ρ_m is rapidly falling to large negative values. Neither of these extremes is physically acceptable. In between with only these data there is a broad range of

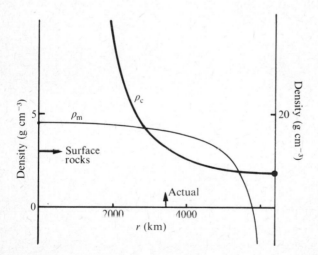

FIG. 2.4. Egg-model values for mantle density, ρ_m and core density, ρ_c for given core radius r. Assumes incompressible material and given mass M and moment of inertia C. Note that the scales for ρ_m and ρ_c are drawn in the ratio $1:4$.

acceptable values. For example, with $r = 3500\,\text{km}$ (close to the actual value of $3475\,\text{km}$) $\rho_m = 4{\cdot}15\,\text{g cm}^{-3}$ and $\rho_c = 12{\cdot}38\,\text{g cm}^{-3}$. Neither of these values corresponds even roughly to densities of obvious materials, but perhaps that is not surprising since we have not yet considered the material as being compressible. We can deduce, however, that the crustal, mantle, and core materials are of three distinct kinds.

3. Jumping bean

SO FAR we have considered only the massiveness of the earth. Now we consider its hardness.

The earth as a bell: mechanical properties at high frequencies

If we place on the ground a seismometer we can directly detect the movement of the ground. Regardless of the location on the earth we find that the ground is in continual motion. It is extremely interesting to record the seismometer signal on a magnetic tape and play back the tape, say, 10 000 times faster. We can then, as it were, listen to the earth. It is a strange experience. Generally it sounds like being in a forest on a windy day, with occasional brief falling tones and longer rather melodic tones, reminiscent of an orchestra tuning up. Every now and again there are sharp noises that sound like a branch breaking. On rare occasions there are sounds like a herd of animals stampeding through a forest, smashing off branches and breaking them underfoot.

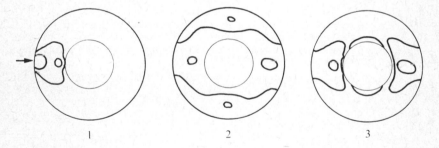

FIG. 3.1. Vibration model of torsional oscillations made from a composite disc observed in the laboratory. Disc struck at one point, marked by the arrow at time zero. Pictures at times 1, 2, and 3 show one cycle of the vibration.

The mechanism for initiating these vibrations is not our immediate concern. What is of first importance is that the earth vibrates at all, how these vibrations are transmitted through the earth, and what they reveal about the nature of the material of the earth's interior and the over-all structural arrangement of that material.

If the earth is thumped hard enough it rings like a bell. This is especially noticeable after a moderately large earthquake or an underground nuclear-bomb explosion. The tones are detected on sensitive long-period seis-

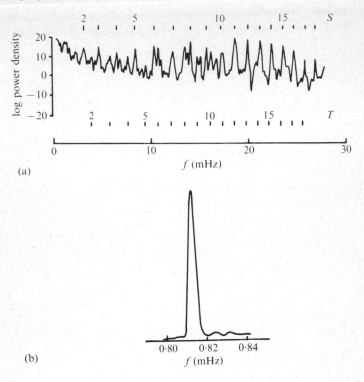

(a)

(b)

FIG. 3.2. Free vibration spectrum, log power density as a function of frequency f, in mHz (1 Hz = 1 c.p.s.), excited by the earthquake of 22 May 1960, Chile, observed at angular distance 44°. (a) Frequency band covering the first 18 modes, for both types S, T; (b) Detail of the very sharp $_0S_0$ peak at 0·81 mHz, a period of 20·5 min.

mometers and reveal a wide range of tones of frequencies down to about 1 cycle per hour. These vibrations of solid bodies are just like sound waves, a regular moving pattern of compressions and torsions accompanied by very small elastic distortions of the material through which they pass. When such waves are confined or accumulated in a volume, certain tones are more persistent than others. This phenomenon of resonance is a common experience; it is exploited in all musical instruments.

Consider, as a rather extreme model of a vibrating earth, a long thin rod. Suppose that waves of compression travel along the rod with velocity α and waves of twisting with velocity β. The condition for resonance of a wave of velocity c is simply that the wave must travel along the rod and back again in precise multiples of the resonant period. Thus, if the length of the rod is l, the period of oscillation T, and the number of vibrations in one full circuit of the wave is n, then $nTc = 2l$.

For compressional waves our model can be compared with vibrations in the earth that are purely radial, the so-called $_mS_0$ modes for which $l = a$ and $n = m+1$.

m	Observed period (s)	nT
0	1228	1228
1	604	1208
2	399	1197
3	303	1212
etc.		

The product nT is indeed nearly constant. Setting $a = 6371\,\text{km}$ we obtain $\alpha = 10{\cdot}4\,\text{km s}^{-1}$, a representative compressional velocity for the earth as a whole.

For torsional waves our model can be compared with vibrations in the earth that correspond to twisting about a diameter, the so-called $_0T_n$ modes for which $l = 2a$ (the $_0T_0$ mode corresponds to a rigid body rotation and is not seismically observable):

n	Observed period (s)	nT
2	2641	5282
3	1708	5124
4	1304	5216
etc.		

Hence $\beta = 4{\cdot}8\,\text{km s}^{-1}$.

Earthquake and explosion seismology

Let us now consider those short sharp sounds that are heard on a speeded-up playback of the earth vibrations. Inspection of a graphical representation of the seismometer displacement shows a distinct but relatively low-amplitude *primary* event P, which starts abruptly and is followed by a period of oscillatory motion until the onset of a generally stronger *secondary* event S, again followed by oscillatory motion until the onset of very much larger, deeply modulated oscillation G which persists for a long time before dying away. If several such records are obtained we notice that the time-intervals PS, SG are roughly in a constant ratio; if PS is large, so is SG. If then records from several stations at different sites are obtained we find that all or nearly all the sites are excited at very roughly the same time. A set of such records can be ordered on the time-intervals PS. From numerous events sets of so-called travel–time curves for P, S are gradually built up. These relations represent the time for seismic pulses to travel from the source region to the observation site.

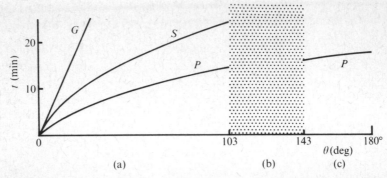

FIG. 3.3. Time t, in minutes, for P, S, and G waves to reach observer at angular distance θ, in degrees, from source. The shaded area represents the region in which waves are diffracted around the core at the base of the mantle.

We notice several things:

(1) P-wave velocities are apparently about $8\,\mathrm{km\,s^{-1}}$ near the source and increase with distance from the source;

(2) S-wave velocities are approximately half P-wave velocities;

(3) in region a of the travel–time diagram both P and S are present;

(4) in region b, $103° < \theta < 143°$, both P and S are absent. This is the so-called *shadow zone*;

(5) in region c only P is present;

(6) the G events have constant apparent velocity.

Laboratory studies of vibrations in solid material show that there are two types of wave: compressional and torsional. In waves of compression, just as with sound waves in a gas, the wave progresses by alternately compressing and dilating the material. In waves of torsion the material is alternately twisted one way and then the other without change of volume. Torsional waves are of lower velocity than compressional waves; and they cannot be generated in a fluid.

If at the observation site all three components of the ground motion are measured, namely, the vertical and two mutually perpendicular horizontal components, then P events can be identified as compressional waves and S events as torsional.

We can make the following deductions:

(1) the G events are surface waves localized in the outer part of the earth;

(2) P and S waves travel through the body of the earth;

(3) immediately below its surface the earth is solid, but at some depth, corresponding to the edge of the shadow zone at $\theta = 103°$ and the complete absence of S waves beyond that, the earth is fluid;

(4) because travel–time curves derived at different sites are nearly identical, the earth is nearly radially symmetric.

Theoretical sketch: core-radius model

We can deduce the core radius from the observed times: the so-called *PcP* at 0° of 8 min, the wave directly reflected at the mantle–core boundary back to the observer; the so-called *PKP* at 180° of 20 min, the wave passing directly through the centre to an observer on the other side of the earth. These times are related to the average compressional wave velocity α by

$$t_c = 2(a-r)/\alpha \quad (PcP), \qquad t_K = 2a/\alpha \quad (PKP).$$

Hence $r = \frac{1}{2}\alpha(t_K - t_c) \approx 3740\,\text{km}$, which is quite a good estimate; the actual value is 3475 km.

If we take $r = 3475\,\text{km}$, the two times can be used to obtain an estimate of α separately for the mantle and the core:

$$\alpha_m = 2(a-r)/t_c \approx 12 \cdot 0\,\text{km s}^{-1}, \qquad \alpha_c = 2r/(t_K - t_c) \approx 9 \cdot 6\,\text{km s}^{-1}.$$

The average velocity is thus higher in the mantle than in the core, but both velocities are of the order of $10\,\text{km s}^{-1}$.

Representation of elastic behaviour

For a wide range of materials being mechanically distorted, laboratory measurements show that the stress is linearly related to the strain in the manner:

$$\mathbf{p} = \lambda\theta\mathbf{I} + 2\mu\mathbf{e},$$

where **p**, **e**, and **I** are the stress, strain, and unit tensors. θ is the dilatation (the proportional decrease in volume), and λ and μ are the so-called Lamé elastic constants. Thus the mechanical behaviour of an elastic body can be fully described in terms of its density ρ and *two* parameters representing the elastic properties. In practice, certain combinations are found more convenient than λ, μ. This is all right provided that we remember to use two independent parameters.

1. An all-over pressure change dp, leading solely to a dilatation dθ, defines the bulk modulus or incompressibility $K = \text{d}p/\text{d}\theta$. Sometimes, as here, $\chi \equiv 1/K$, the compressibility is used.

2. A longitudinal stress p in one direction, leading to a stretching (or compression) e in that direction and a corresponding compression (or stretching) e' in the direction at right-angles, defines Young's modulus $E = p/e$, and Poisson's ratio $\sigma = -e'/e$.

3. Seismic waves of compression and twisting, the so-called P and S waves of seismology, define velocities α and β.

A useful approximation for the earth is $\sigma = \frac{1}{3}$ when

$$\beta = \tfrac{1}{2}\alpha. \qquad K/\rho = \alpha^2 - \tfrac{4}{3}\beta^2 = \tfrac{2}{3}\alpha^2, \qquad E = 3(1-2\sigma)K = K.$$

In this case the mechanical behaviour is given by ρ and only *one* elastic parameter.

Thus, for $\alpha = 10\cdot4\,\mathrm{km\,s^{-1}}$, $E \sim 4\,\mathrm{Mbar}$—a material tougher than steel!

Variation of properties with depth

The results of vibration and earthquake seismology when combined with those of geodesy can be manipulated numerically to give estimates reliable to about ± 1 per cent of the average elastic properties of a uniformly stratified earth.

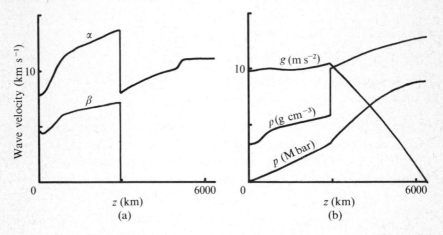

FIG. 3.4. Property variation with depth z, in km. (a) Wave velocities α and β (in $\mathrm{km\,s^{-1}}$); (b) density ρ (in $\mathrm{g\,cm^{-3}}$); pressure p (in Mbar); gravity g (in $\mathrm{m\,s^{-2}}$). Plotted values of p are $2\cdot5 \times p$.

Implicit in these deductions are strong assumptions about the equation of state of the rock-substance which constitutes the interior. One simple possibility is

$$\rho = \rho_0(1 + n\chi_0 P)^{1/n},$$

where ρ is the density at pressure P, ρ_0 the density at zero pressure, and n and χ_0 are constants for a given rock-substance. We notice that the incompressibility $\chi \equiv (\partial\rho/\partial P)/\rho$ satisfies

$$\chi\rho^n = \chi_0\rho_0^n = C \text{ (a constant)}.$$

If we were interested in the fine detail, the earth could be considered as a two-layer mantle plus core. In this case we note that the densities of the three major rock-substances at zero pressure are $3\cdot29\,\mathrm{g\,cm^{-3}}$, $4\cdot09\,\mathrm{g\,cm^{-3}}$,

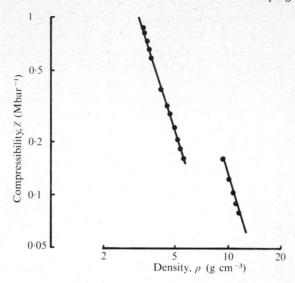

FIG. 3.5. Compressibility χ, in $Mbar^{-1}$, as a function of density ρ in $g\,cm^{-3}$. Lines of slope $n(=3)$ are drawn and show $\chi\rho^n \approx C$ (a constant).

and $6\cdot24\,g\,cm^{-3}$—the last being the zero-pressure density of the core liquid. Thus the upper mantle in this model has the density of an ultrabasic rock (actually this is built into the model); the others are unknown. But it is worth noting that the core material could not be a liquid iron–nickel mixture as is so often stated, for then its (liquid) density would need to be not less than $7\,g\,cm^{-3}$.

For most purposes a single-layer mantle is adequate. In calculations done for this book $n = 3$ is used throughout.

TABLE 3.1

Properties of an earth with a two-layer mantle

	Top	Bottom	Core
Depth of base (km)	550	2900	6370
ρ_0 $(g\,cm^{-3})$	3·29	4·09	6·24
χ_0 $(Mbar^{-1})$	0·904	0·440	0·646
n	4·35	3·34	3·40
C $(Mbar^{-1})$	160	49	326
E_0 (Mbar)	1·5	2·0	—

Viscous drop: mechanical properties at low frequencies

The most direct method of obtaining information about the distribution of mass inside the earth is by measuring the external gravity field. This is normally done with a gravimeter, which is simply an elaborate spring balance. This instrument measures the variation of the vertical component of gravity on a fixed mass. Thus, for the sake of illustration, consider a roughly spherical mass of radius 1 km and density $2 \cdot 7 \, \mathrm{g \, cm^{-3}}$ buried in ground of density $2 \cdot 4 \, \mathrm{g \, cm^{-3}}$ with its centre at depth 2 km. The vertical component of gravity, measured at the surface on the vertical through the centre, is $g' = 2 \cdot 1$ mgal (mgal = milligal; 1 gal = $1 \, \mathrm{cm \, s^{-2}}$). Compared with the total field of about 10^6 mgal this is very small.

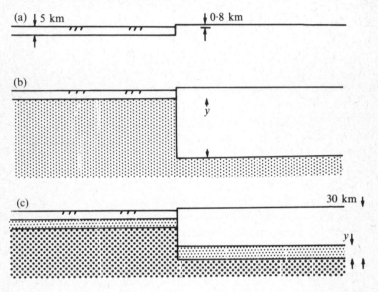

FIG. 3.6. Models of the crust: a 200-km transverse section near a hypothetical continental margin: (a), (b), and (c) are progressively more realistic. Densities (in $\mathrm{g \, cm^{-3}}$); unshaded 2·7; light 3·0; heavy 3·3; hatched 1·0 (water). (a) Homogeneous crust, 5 km ocean depth, 0·8 km mean land elevation. (b) Basaltic oceanic crust and mantle, granitic continent of thickness $(5 \cdot 8 + y)$. (c) Distinct oceanic crust, 5 km thick; continental crust of total thickness 30 km, upper part granitic, lower part basaltic of thickness y in a homogeneous mantle.

Now let us see what the situation is near a continental margin, a place of obvious and extreme change in the topography of the crust. Suppose we assume that the crust is more-or-less homogeneous; then the margin will be as shown in model (a). The gravity produced by a slab of thickness h and density ρ is $g' = 2\pi G \rho h$. Notice incidentally that this is independent of the depth of burial but applies only at points distant from the ends of the slab. We have 41·8 mgal per km $\mathrm{g \, cm^{-3}}$. Thus, in model (a), the difference between gravity over the land and over the sea is about $(5 \cdot 8 \times 2 \cdot 7 - 5 \times 1) \times 41 \cdot 8 \approx 450$ mgal. We set out to measure this, and find that the difference is in fact very much smaller. Apart from certain small (but important) areas a closer approximation would be to say, as a first approximation, that gravity variations are negligible.

Our first model of the crust was a hopeless failure. We can now improve it by recognizing that in fact the crust is not homogeneous but that continents are granitic and ocean basins basaltic. Thus, if we now consider model (b) with the granitic crust as just a patch of material imbedded to a depth y in an otherwise basaltic earth and require that the variation of gravity is precisely zero, then $5 \times 1 + 3y = 2 \cdot 7(5 \cdot 8 + y)$.

This gives $y = 35 \cdot 6$ and the thickness of the continent is $41 \cdot 4$ km. At this stage we must do some explosion seismology to determine directly the presence of our postulated granitic–basaltic boundary at the base of the crust.

This model is not quite the disaster of the first one. We do find a reflector beneath the continent, but it is generally only about 30–35 km deep, and the boundary seems to carry on into the oceanic area at a depth below the ocean floor of only 5 km. This boundary is the so-called Moho. Moreover, the seismic velocities suggest that the lower continental crust is seismically similar to the oceanic crust and the entire crust lies on a more dense material, the mantle, of density $3 \cdot 3$ g cm^{-3}.

Taking all this into account in our third model (c), the remaining unknown is the thickness y of the lower continental crust. As before, equating the two gravity contributions:

$$5 + 5 \times 3 + 3 \cdot 3(30 - 10 - 0 \cdot 8) = 2 \cdot 7(30 - y) + 3y,$$

we have $y = 8$ km. Seismic observations of the layering of the continental crust are not very clear-cut. Nevertheless, numerous identifications of such layering have been made; for example, the lower crust in Europe is about 5–10 km thick.

We could go on. But this is enough. Our method of expressing the normally small gravity effect of the crust can be put another way. We have simply been totalling the contributions ρh for each layer. But this is just a way of finding the total pressure at a given depth. If the pressures turn out to be equal there will be no tendency for relative movement. This does not mean that pressures are identical at every level, but only at and below the bottom of the continental crust. We imply in this case that the continental mass is stiffer than its surroundings. But the general point is that the crust is floating in the mantle.

The geoid

Recent measurements of the motions of artificial satellites have defined the global gravity pattern far more extensively and accurately than before. The equipotential surface, which over the oceans approximates to mean sea-level and is called the *geoid*, is a bumpy surface with variations in height of generally less than ± 100 m. The corresponding gravity variation is ± 10 mgal. Further, these bumps are not correlated with the distribution of oceans and continents nor with any of the large-scale features of the crust.

Noting again that the observed values of gravity are very much less than those given by the apparent excess land mass, we deduce (1) that the crust floats in the mantle and (2) that the bumps on the geoid arise from mass anomalies deep in the earth.

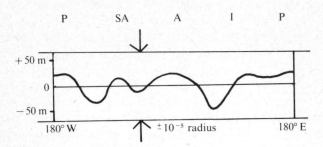

FIG. 3.7. Variation of the mean form around the equator for bumps greater than 2500 km across. P, Pacific; SA, South America; A, Africa; I, Indonesia.

The ultimate test of the floating idea is, however, to place a load on the crust and see if the crust sinks deeper in a more dense mantle far enough to produce the additional buoyancy required, and then to remove the load and see if the crust floats up again to its old level. And Nature is doing this all the time!

The floating crust

One of the intriguing features of the earth's surfaces is the frequent and persistent undulations. The surface elevation may go up and down by about 1 km over extensive areas for various periods of time. The elastic movements (such as seismic waves) are minute, and very rapid compared with these relatively slower undulations: however some undulations are particularly rapid.

The most notable is the present-day upward movement of the so-called Fennoscandia region, centred on Norway, Sweden, and Finland. Observations of raised beaches and tide-gauge records show an elevation of about 300 m in the past 10^4 years. Compared to most other geological processes this is extremely rapid. The present view is that this region of the earth's surface was depressed under the weight of the ice cap of the last ice age and is now returning to its former position.

We can model such rapid undulatory movements by considering the earth as a viscous body on which an undulation has been imposed, and inquiring into its subsequent behaviour. An imposed undulation of wavelength λ will decay in a time proportional to τ, where $\tau = 4\pi v/g\lambda$, v is the kinematic viscosity (see the glossary), and g the acceleration due to gravity. Thus, taking data roughly appropriate to Fennoscandia, namely, $\lambda = 1500$ km,

$\tau = 10^4\,\text{yr}$, we have $v \sim 3 \times 10^{16}\,\text{m}^2\,\text{s}^{-1}$. This corresponds to a viscosity, $\mu = \rho v$, with $\rho = 3 \cdot 3\,\text{g}\,\text{cm}^{-3}$, of about $10^{21}\,\text{P}$ (P = poise).

Theoretical sketch: isostatic movement

A simple description of the rate at which objects floating in the mantle return to their equilibrium position is given by the following model.

1. We recognize that crustal material is extremely viscous and that the viscosity of the mantle decreases rapidly with depth because of increasing temperature. Thus we consider a thin, flexible, but otherwise rigid, crust floating in a viscous upper mantle which itself lies above a relatively inviscid lower mantle.

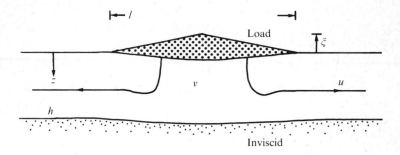

FIG. 3.8. Diagram for model of flow produced by a crustal load on a viscous layer above a deep relatively inviscid layer.

2. If a load is placed on the crust over an area of width l, much greater than the thickness of the upper mantle h, the dominant flow process will be the horizontal flow in the upper mantle channel.

3. Thus the horizontal velocity $u(z) = p'z(2h-z)/2\mu$, for which on the base of the crust ($z = 0$) we have $u(0) = 0$, since the crust is rigid; on $z = h$, the base of the upper mantle, the shear stress is zero, since the lower mantle is inviscid; and p' is the horizontal pressure-gradient. Hence the horizontal discharge $Q - \frac{2}{3}p'h^3$. But this discharge is produced by the fluid displaced by the vertical movement of the crust. Taking an average vertical velocity w for this movement, $Q = wl$.

4. Finally we need to inquire into the source of the horizontal pressure gradient. Let us assume that the crust has been buried under a load equivalent to a maximum height ξ of rock, the thickness of the superimposed layer diminishing uniformly from the centre to the edge. Thus, $p' = 2\rho g\xi/l$.

5. Noting that $w = -\frac{1}{2}d\xi/dt$, then $d\xi/dt = -\xi/\tau$, where $\tau = 3vl^2/8gh^3$ is the exponential time-constant of the motion.

6. For $\tau = 10^4\,\text{yr}$, $l = 10^3\,\text{km}$, $h = 10^2\,\text{km}$ we estimate $v \sim 10^{16}\,\text{m}^2\,\text{s}^{-1}$. This value is close enough to the previous estimate to give us confidence in the model.

The chief new point emphasized by this model is that for a mantle with viscosity decreasing with depth the method of estimating viscosity from vertical displacement of portions of the crust gives a value not for the earth as a whole but for that part of the mantle immediately below the crust. The ultimate behaviour is determined by the most viscous material in the immediate vicinity of the loaded crust—simply because that is the region which will take longest to adjust because of its higher viscosity.

(A somewhat more refined model, for which $\mu = \mu_0 \exp\{(h-z)/\delta\}$, gives $\tau \sim v_0\, l^2/8g\delta^3$ for $\delta \ll h$, where δ is a measure of the thickness of the zone in which the viscosity is important.)

4. Rock-substance

The rheology of the working substance

THE EARTH behaves elastically at high frequencies and viscously at low frequencies. How is this possible? By now everyone must have played with 'silly putty', which can be smashed with a hammer blow, bounces like a ball, and can be pulled out like chewing-gum.

FIG. 4.1. Flow of the solid rock-substance ice directly observed. The glacier Quvnertussup Sermia at 53° W, 71°15′ N. Width of flow about 1 km. (By courtesy of the Geodetic Institute, Copenhagen, part of aerial photograph A77/112: 20.)

A perfectly elastic solid is a conceptual idealization. Real materials have structural 'imperfections' which prevent perfectly elastic behaviour. Even for stresses low enough to produce no permanent deformation the total strain is made up of two parts: an elastic part proportional to load, and a time-dependent, fully recoverable part that varies with rate and duration of loading. This anelastic behaviour can be described in terms of rearrangement of the microstructure of the material. Anelastic behaviour is not restricted to small deformations, but when a material deforms permanently the anelastic effects are engulfed in the plastic behaviour.

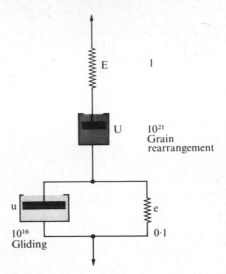

FIG. 4.2. Spring and dashpot model of a viscoelastic material. E, e represent purely elastic springs; U, u represent purely viscous effects; connected together by rigid rods. Typical magnitudes are shown for E, e in Mbar, for U, u in P.

A mechanical model of springs and dashpots can be used to represent these properties qualitatively.

(1) At low strain, if the 'viscosities' of (U, u) are such that $U \gg u$; the anelastic effects are such that the 'elasticity' of (E) controls the purely elastic strain; and (e, u) together control the time-dependent strain.

(2) If the model is elongated slowly, u being small barely resists the extension in e, and the elasticity will arise from the springs (E, e) in series.

(3) In rapid elongation, (u) will hardly have time to move, so that (e, u) will act as if it were rigid and the elasticity will be that of spring (E) acting alone.

(4) If the model is loaded slowly and unloaded rapidly the strain will not immediately return to zero. What appears to be a permanent strain will be observed. The strain will return to zero when the stress stored in (e) and trapped by (u) has been relaxed.

(5) If a mass is attached to the bottom of the model and made to vibrate freely up and down, the amplitude of vibration will decrease with each cycle owing to the dissipation of energy in (u).

(6) Permanent loading, after an initial elastic adjustment of (E, e), will be controlled solely by (U), when the strain will increase linearly with time—purely viscous behaviour.

The effect of rearrangement of the microstructure on the macroscopic properties of a substance can be demonstrated in many ways. A very nice example is to fill a bladder as full as possible with sand completely saturated with deaerated water and tightly sealed. Normally the bladder is quite squashy and soft, but if two sides of the bladder are sharply jerked apart the bladder immediately becomes stiff and hard. Here the altered packing of the sand grains has changed the void space, and since water is incompressible a large pressure change occurs in the water. An intriguing extension replaces part or all of the sand with salt crystals and the water with partially or completely saturated salt solution. The phenomenon is reproduced with the additional effect that there is a slow return to the soft state owing to progressive solution and recrystallization. If you use a child's balloon and table salt you will find that the return takes about an hour.

On the weakness of rock

A simple dimensional analysis shows that materials corresponding in the laboratory to an earth considerably stronger than steel are still extremely weak. For long-term geological processes stresses must be balanced by the body force of gravity. Consequently a key measure of the strength will be $\xi = S/g\rho L$, where S is the breaking stress, ρ is the density, and L is a characteristic length.

Thus in modelling a geological object we need to make the ratio the same in the model as in the object.

Consider, for example, a crustal slab of depth $L = 50$ km of a material with a breaking strength 4 kbar; this is actually the strength of a typical steel. Let us make our model 20 cm thick. We then find that the material must have a strength $S = 0.016$ bar. This is an extremely weak material—weaker than butter. At, for example, a model density of 1 g cm^{-3}, a 0.2 m column of such material could not sustain its own weight: the stress at the base would be 0.02 bar.

Now consider a self-gravitating sphere modelling the entire earth with the constant of gravitation altered so as to keep the surface acceleration of gravity the same as for the earth. For an earth with the strength of steel, a 1-m sphere should be made of very soft mud with a strength of 3×10^{-4} bar.

The argument here is similar to that used by Galileo in his *Discourses concerning two new sciences* (1638) to demonstrate the impossibility of indefinitely large animals. Total weights increase as the cube of a linear dimension, but the supporting power of the skeleton increases only as the square.

The flow of crystalline solids

In the simplest case, relative movement of parts of a single crystal occurs by the slipping of a pair of adjacent planes of atoms past one another, the relative displacement being parallel to the slip-plane. All the atoms on either side of the slip plane do not necessarily move simultaneously. At any given moment after the application of the stress, some of the atoms on the plane will be already displaced; others will not. These patches in the slip-plane in which displacement has already taken place will be separated from the other portions of the plane by lines, generally simple closed curves. Near these lines the atoms will be severely displaced from their accustomed relation to the crystal lattice. Hence these lines are called dislocations. Thus we envisage the deformation of a single crystal as arising from the generation and movement of dislocations, notably along slip-planes. Various types of dislocations have been directly observed and much is known about them.

When a material is well below its melting temperature, the bulk of the dislocations appear under the action of an applied stress. (Some dislocations are, of course, always present.) But when the temperature is comparable with the melting temperature T_m—typically, greater than $0.4\,T_m$—there are numerous dislocations present and it is more appropriate to think of these as diffusing through the material.

Direct measurements of the flow of single crystals of ice give a viscosity about 2×10^{11} P, whereas the viscosity of natural polycrystalline ice gives about 10^{14} P. Clearly the movement of dislocations is greatly inhibited by the grain boundaries and the flow is largely determined by the much slower process of the rearrangement of grain boundaries. We can expect a similar situation with any solid rock-substance.

But this is by no means the end of our difficulties. In all but the smallest-scale studies the basic component of a geological system is what I shall call a *rock unit*. This unit is a volume of mixed rocks, made, for example, of a sequence of layers as in, say, a series of lavas and interbedded sediments. Thus whereas the petrologist regards a rock as a crystalline mixture, a rock unit is a rock mixture.

The obvious difference between ice and other rocks is that whereas ice is monomineralic, silicate rocks are generally polymineralic. Clearly it is in general more difficult to move grain boundaries in a polymineralic substance. It should be no surprise therefore that silicate rocks are several orders of magnitude more viscous than ice.

Nevertheless it is not possible at the present time to deduce the viscosity of a rock unit from a knowledge of its composition and structure. We have no alternative to field measurement. Take, for example, the case of a rock unit flowing down a slope. The average velocity of the flow is $\rho g \sin \theta\, h^2/3\mu$, where θ is the angle of the slope to the horizontal and h is the depth of the rock unit. For example, a glacier of thickness 100 m on a 2° slope flows

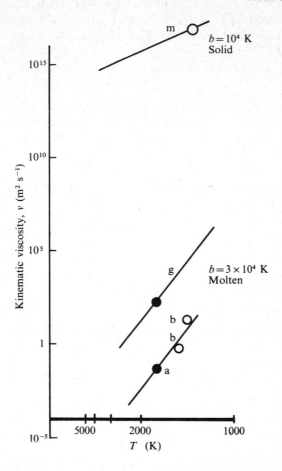

FIG. 4.3. Kinematic viscosity, v in $m^2 s^{-1}$ of various solid and liquid rock-substances as a function of absolute temperature T, in K where v is plotted on a logarithmic scale against $1/T$. A straight line on this plot corresponds to $v = v_0 \exp(b/T)$, where v_0 and b are constants. The two lines at the bottom are for molten material. m, mantle; g, granite; b, basalt; a, andesite.

4 m in a year, giving $\mu \sim 10^{14}$ P; about the same value as directly determined for the polycrystalline ice. On the other hand, a flow nappe on a 6° slope could have travelled 3 km in a million years, giving $\mu \sim 10^{20}$ P.

Viscosity variation with temperature

The rate of diffusion of dislocations and rearrangement of grain boundaries will clearly increase rapidly with temperature. This is especially pronounced in its effect on the viscosity.

It is common experience that viscosity decreases rapidly with temperature.

For most substances $v = v_0 \exp(b/T)$, where T is the absolute temperature and b is a constant. Surprisingly little has been done to determine b for rock-substance, either molten or solid. The data shown, though based on some measurements, are just informed guesswork. The essential features to notice are:

(1) the viscosity of the molten material is about 10^{-15} that of the solid;
(2) granitic melts are about 10^3 times stiffer than basaltic melts;
(3) the parameter b is less for a solid than for a melt (based on calculations like those in Chapter 9);
(4) as for liquids, it is assumed here that acidic solids are substantially more viscous than basic solids, and this is undoubtedly the basis of the persistence of portions of the granitic crust.

To my mind one of the most convincing pieces of evidence that the viscosity decreases with temperature and hence with depth in the earth comes from the distribution of earthquake foci. Earthquakes seem to be of two kinds: shallow earthquakes, which occur usually in the top 20 km or so and are analogous to brittle fracture; and deep earthquakes, which occur usually at depths of the order of 100 km and are more like implosions, possibly arising from rapid local phase changes. But these do not occur below about 700 km. The material is just not stiff enough.

5. Reactor

SO FAR we have considered the earth simply as a massive object—a mere lump of matter. But to make anything happen we need both matter and some energy.

Here I want to consider the earth simply as a machine for processing rock. This viewpoint is rather like that of a chemical engineer designing a large chemical plant, with the important difference that the actual machinery is not fixed and can itself also be involved in the processes. Hence we need to assess three major constituents: the working materials, the chemical processes, and the dynamical processes.

During the course of the process matter is transformed by drawing on a supply of energy; this matter and energy are the working materials.

The matter available is to a close approximation that of the present earth. The gain of mass due to meteoric material, etc. or loss due to the escape of volatile gases during geological time is a minute proportion of the earth's mass. Notions that great chunks of the earth have been removed can be completely discounted.

The crust and the upper mantle provide a reservoir of rock from which material is withdrawn. Yet we know that the suites of basic igneous rocks have been remarkably the same throughout geological time. This would not be possible from a local closed reservoir, since the constituents for particular rocks would progressively be exhausted. It is therefore necessary to consider mixing of upper and lower mantle material.

The energy sources available are gravitational energy, thermal energy, chemical energy, and nuclear energy.

There is no doubt that the interior of the earth is hot. For example, in parts of Kyushu, Japan, if you take a hollow bamboo pole and hammer it a metre or so into the ground, after a short time steam will come out of the top of the pole. This steam is used by the villagers. At Wairakei, New Zealand, concrete and iron pipes 1 km deep discharge in total several tons per second of hot water and steam for an electric power station. But nature has made her own even larger pipes: sheets of ignimbrite, produced in a single eruption, cover several thousands of square kilometres; the Tibesti volcanic complex is like a porridge pot 300 km wide. Need I go on? Clearly we have a problem in heat and mass transfer. But where do the heat and mass come from, and how are they transported?

Before we look into the details of these processes it is necessary to consider how the various parts of the earth arranged themselves as they are. Thus we

Fig. 5.1. Steam emerging from hollow bamboo poles rammed into the ground. Kyushu, Japan.

consider first the over-all nature of the jam-pot itself, avoiding anything but the broadest considerations. For the moment our jam-pot is thought of as a blast-furnace.

Surface heat flow

The simplest way to determine the heat transferred at the surface is to drill a hole in the crust, measure the temperatures down the hole, and measure the thermal conductivity of a sample of the rock. There is a surface zone, extending down to about 100 m, in which temperatures are variable because of diurnal heating and movement of water and water vapour. But below this zone almost everywhere there is a steady increase of temperature with depth, typically $25 \, \text{K km}^{-1}$ in a region of thermal conductivity K of, say, $2 \, \text{W m}^{-1} \, \text{K s}^{-1}$. The heat flux f and the temperature gradient $\text{d}T/\text{d}z$ for heat transferred by thermal conduction are related by

$$f = K \, \text{d}T/\text{d}z.$$

This in effect is the definition of K. Thus for the above figures $f = 50 \, \text{mW m}^{-2}$. A square of ground 150 m across produces about 1 kW.

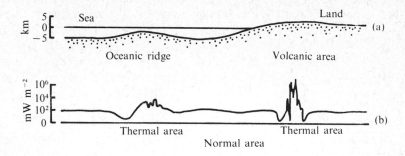

FIG. 5.2. Schematic profile of net outward surface heat flux: (a) topography; (b) heat flux.

Two distinct regions can be recognized.

1. Normal areas, covering more than 99 per cent of the earth's surface, are those in which the heat flux lies in the range 0–$100\,mW\,m^{-2}$, with an average of about $50\,mW\,m^{-2}$. The vertical temperature gradient is nearly constant to depths exceeding $1\,km$, and the surface heat flux variations are gradual, usually being negligible over distances of the order of $1\,km$.

2. Thermal areas are those in which the heat flux can reach $1\,kW\,m^{-2}$, though it is generally much smaller, with possible average values of the order of $1\,W\,m^{-2}$ over areas of the order of $10^3\,km^2$. Large horizontal variations of heat flux and vertical variations of vertical temperature gradient are possible over distances of $1\,m$.

Two extremely important results emerge from a global study of the surface heat transfer.

1. There is everywhere over the earth's surface a *net* outward flow of heat. Heat is being continually brought to the earth's surface from its interior and lost.

2. Apart from certain narrow areas, to be discussed later, the heat flux is nearly uniform over the surface. In particular it is the same over continental platforms as over oceanic basins.

Temperatures of the interior

Clearly a temperature gradient of $25\,K\,km^{-1}$ cannot be uniform throughout the interior: at $1000\,km$ the temperature would be $2.5 \times 10^4\,K$, enough to vaporize the material, whereas it is a solid. Although it is little more than a guess, the temperature at the base of the mantle could be, say, $2500\,°C$, the approximate melting temperature to be expected for silicates at pressures of the order of 1 Mbar. If the mantle is *well mixed* the temperature will closely approximate to adiabatic so that $T = T_0 \exp(\gamma g z/c)$, giving a drop to about 1800 °C at the surface. Between these two lines we can draw the melting-point line for basalt, a ubiquitous product of the mantle with a surface temperature of 1100 °C and melting-point gradient about $2\,\mathrm{K\,km^{-1}}$. Basalts seem to originate at depths of up to 50–100 km, and over this range

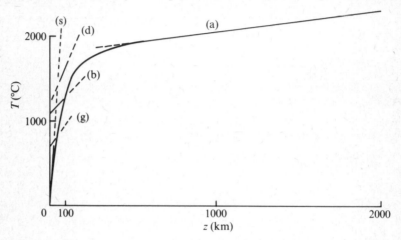

FIG. 5.3. Temperature of the interior. Mean temperature T (°C) as a function of depth below the surface z (km). A possible profile is shown (solid line); based on various criteria: (a) deep interior well mixed, adiabatic (isentropic); (s) surface gradient, here drawn at $25\,\mathrm{K\,km^{-1}}$; (g) melting-point gradient for granite; (b) melting-point gradient for basalt; (d) melting-point gradient for certain mineral mixtures, notably garnet–lherzolite, from kimberlites. This diagram also emphasizes our ignorance of detail below about 100 km.

of depth the temperature should therefore be near the basalt melting temperature. Similarly, if we assume that the thickness of the granitic crust is determined by the melting of granitic rock-substance, with a surface melting temperature when wet of 700 °C and melting-point gradient about $4\,\mathrm{K\,km^{-1}}$, below 30–40 km temperatures should exceed the granite melting temperature. There is a further suggestion from studies of kimberlites. Certain pyroxenes found in the ultrabasic nodules have a surface melting temperature of 1200 °C and melting-point gradient about $4.5\,\mathrm{K\,km^{-1}}$. There is also the notable but rare occurrence of diamond, which indicates brief temperature pulses up to 2000 °C at 100 km or so. Other than this we have no further direct information.

Nevertheless it is immediately apparent that the internal temperature distribution has two properties:

(1) in a surface zone to depths of the order of 100 km the mean radial temperature gradient is large and strongly superadiabatic;

(2) below the surface zone mean radial temperature gradients are small, about $0.2\,\mathrm{K\,km^{-1}}$, and within a factor of 2 are close to adiabatic.

Geological energy consumption

If the net heat flux throughout geological time had been constant at $50\,\mathrm{mW\,m^{-2}}$, the total loss of energy from the earth's interior would be $4 \times 10^{30}\,\mathrm{J}$. This is equivalent to the energy lost from a body of the thermal capacity of the earth, about $5 \times 10^{27}\,\mathrm{J\,K^{-1}}$, for an average fall of temperature of $670\,\mathrm{K}$, a substantial portion of the earth's initial thermal energy. For example, if the earth's initial mean temperature were, say, $3000\,^\circ\mathrm{C}$ this is about 27 per cent of the initial internal energy. Our concern now is to find out how to get our hands on all this energy, or rather how to obtain access to it.

Energy required for global crustal movement

At first sight it seems as if enormous amounts of energy and large stresses would be required to work the global machinery. This is not so. Consider a crustal slab and an attached portion of mantle of depth h moving with velocity U. The viscous stress is $\tau \sim \mu U/h$ and the rate of working per unit area $e \sim \tau U$. For, say, $U = 3\,\mathrm{cm\,yr^{-1}}$, $h = 10^2\,\mathrm{km}$, $\mu = 10^{21}\,\mathrm{P}$, then $\tau \sim 10\,\mathrm{bar}$ and $e \sim 1\,\mathrm{mW\,m^{-2}}$. In other words, in this case only about 2 per cent of the global heat flux is needed to drive such a slab with stresses equivalent to the pressure of $30\,\mathrm{m}$ of rock.

Gravitational energy of an orogenic belt

For an idealized orogenic belt of area A, elevation h, and thickness of root H, the gravitational energy of the root and the uplift, $U = \frac{1}{2}Ag(H^2\Delta\rho + h^2\rho_\mathrm{c})$, where $\rho_\mathrm{c}, \rho_\mathrm{m}$ are the densities of the crust and mantle and $\Delta\rho = \rho_\mathrm{c} - \rho_\mathrm{m}$. If the belt is also in isostatic equilibrium $h\rho_\mathrm{c} = H\Delta\rho$. For the Alpine–Himalayan belt, roughly $500\,\mathrm{km} \times 10^4\,\mathrm{km}$, of average elevation $2.84\,\mathrm{km}$ and with a root of $20\,\mathrm{km}$, we have (assuming isostasy) $U = 4.5 \times 10^{24}\,\mathrm{J}$. Now if this belt were to be created or eroded away in, say, 45 million years the power required or destroyed would be $10^{17}\,\mathrm{J\,yr^{-1}}$. This at first sight seems a great deal of energy but it is an energy flux of about $0.1\,\mathrm{mW\,m^{-2}}$ as compared with the normal geothermal flux of about $50\,\mathrm{mW\,m^{-2}}$. We can make a lot of mountains with negligible effect on our energy budget.

Inaccessibility of heat for thermal conduction

The inability of thermal conduction to get access to the amount of heat required can be demonstrated in two equivalent ways. Both use the knowledge of how an initially uniformly hot sphere cools down. For a sphere of radius a and thermal diffusivity κ the dimensionless temperature θ is a function solely of position r/a and time t measured in the unit a^2/κ.

1. Suppose for argument's sake we wish to obtain about half the initial

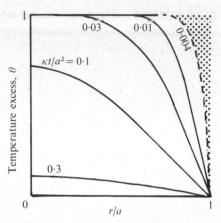

FIG. 5.4. Temperature distributions within a uniformly hot thermally conducting sphere of radius a, suddenly cooled at the surface, at various times t. Values of $\kappa t/a^2$ are shown. θ = dimensionless temperature excess over ambient value. The shaded area shows the amount cooled for a body the size of the earth.

internal energy. The cooling curves shown then require $\kappa t/a^2 \approx 0{\cdot}1$, for which $t \approx 10^{11}$ yr. This is considerably longer than we have already been in business and the surface temperature gradient would be only $0{\cdot}5\,^{\circ}\mathrm{C\,km^{-1}}$.

2. For a geological age of 5×10^9 years the dimensionless time $\kappa t/a^2 = 0{\cdot}004$. As shown, only a very thin layer has been cooled in this time. For example, $\theta = 0{\cdot}5$ at a depth $z/a \approx 0{\cdot}06$.

Thermal conduction alone is clearly a feeble process.

The possible role of radiative transfer

The obvious escape from this difficulty is to find a process with readier access to the internal energy store; indeed, this is the approach of this book. But first we must look briefly at an earlier candidate. At the temperatures envisaged in the earth's interior and if the rock-substance were sufficiently transparent to thermal radiation we could have access to the interior. The contribution of radiative heat transfer can be described quantitatively as a thermal conductivity proportional to T^3/ε, where T is the absolute temperature and ε, the extinction coefficient, is a measure of the opacity of the medium. High hopes were held for this process until laboratory measurements showed that the radiative thermal conductivity was rather small. Rock-substance is not very transparent!

Bring in another energy source: radiogenic heat production

If there is no possibility of getting access to enough of the internal energy perhaps there is an alternative energy source available. And indeed there is! Radioactive elements produce heat as a result of their nuclear decay.

But as we have already seen, there will be no point in burying radioactive elements throughout the earth since only the energy produced in the thin outer layer will be accessible. Just where we put this material requires a little care because there are two important geological facts to consider: (1) the large atoms of the radiogenic material strongly prefer to be in acidic rocks; (2) the values of the heat flux through the continental and oceanic crusts are about the same.

The ultimate absurdity

The essential requirement of all these ideas of purely conduction models is that heat is diffusing through a fixed structure. The structure itself may not move except extremely slowly. If mass movements carry heat from place to place at a rate comparable with or greater than that of thermal conduction all the predictions of the conduction models are invalid. We know from the rates of crustal rearrangement that heat transfer by mass movement reaches values of the order of 10^3 times that given by pure conduction in some parts of the system; we will discuss this in much more detail later. The conduction models fail, not so much because of their sheer inability to give access to and transfer an adequate amount of energy, but because they are quite inappropriate to a heat engine dominated by vigorous mass movements.

There is no alternative to a fully convective model!

Layout of the chemical plant

As a first attempt to design the machinery to process planetary quantities of rock-substance we consider an interconnected collection of vats. In each vat various processes act on the input material and energy to produce products for supply to other vats. This is a continuous flow system using up its built-in energy supply, but retaining the total rock-substance in a variety of forms. The sizes of the vats change with time because of net differences between input and output of matter.

We recognize three types of vat.

1. Entirely fluid: a single such vat the core—large.

2. Largely solid but of generally moderate viscosity: a single such vat the mantle—very large. Perhaps it would be more realistic to treat the mantle as a series of vats, rather like a sort of three-dimensional distillation column, but in this book it is treated as one vat. Nevertheless a very clear and important distinction is made between the upper and lower mantle.

3. Small crustal vats of rather viscous almost entirely solid material. Here I consider three such vats: oceanic, continental, and sedimentary. The sediment vat is the rubbish-dump of the crustal system. Even though much of the sedimentary material is spread over the entire crust we can think of it as being processed in a single vat.

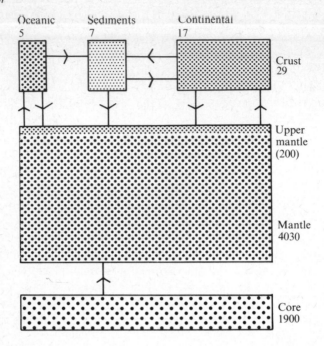

FIG. 5.5. Vat model of the chemical reactor. Units: 10^{21} kg. Areas are drawn proportional to mass, but the three crustal vats are drawn 200 times bigger again: the crustal vats are very small. Masses are for the vats now.

The crustal processes are rapid. In other words, the rates of transfer are such that the mean transit time for a particle of rock-substance through a crustal vat is very small compared with the span of geological time.

The plan for this book is in effect to look in turn at the processes in these vats, concentrating first on the largest one, the mantle.

TABLE 5.1
Masses and dimensions of the main parts of the earth

	Mass	Mean density	Surface area	Radial thickness	Volume	Mean moment of inertia
	(10^{21} kg)	(g cm^{-3})	(10^6 km^2)	(km)	(10^9 km^3)	(10^{35} kg m^2)
Core	1900	11·0	151	3471	175	90
Mantle	4030	4·5	505	2900	898	705
Crust (continental)	17	2·75	242	30	7·3	6
Crust (oceanic)	5	2·9	268	6	1·6	2
Ocean	1·4	1·03	361	3·8	1·4	1

The question of the chemistry

If we wished to learn how to cook an egg, how much would we learn by making measurements on the chemical composition of the water in which it is boiled? Not much, since cooking merely involves elevating the temperature so that chemical reactions inside the egg proceed quickly. In general the ratio of a chemical reaction rate to a dynamical rate is very small. Most silicate reactions are geologically extremely rapid, requiring typically a second to a year to completion, so that to a very close approximation we can assume local chemical equilibrium.

Nevertheless it is very important to realize that this does not imply chemical homogeneity in a macroscopic system. Quite the contrary! The pressure and temperature vary throughout the system and hence so does the chemistry. In addition, for a dynamical system, there is a continuous process of transporting matter from one place to another so that the same piece of matter will pass through varying chemical states in its passage around and through the system.

The bulk of the rock-handling is:

(1) production of basalt for lining new ocean floors;
(2) metamorphism and granitization of sedimentary junk in orogenic systems; and
(3) annealing of mantle material by continual recycling.

Review of energy stocks

Gravitational energy of an egg

The self-attraction of the mass of a planetary body provides an energy source related to the internal mass distribution.

For a sphere of uniform density the gravitational energy per unit mass is $\frac{4}{5}\pi G\rho a^2$. With $\rho = 5.517\,\mathrm{g\,cm^{-3}}$, the earth's mean density, we have $3.75 \times 10^4\,\mathrm{kJ\,kg^{-1}}$. This is an enormous amount of energy but not all of it is available. We have to account for:

(1) the compressibility of rock-substance producing a non-uniform body;
(2) changes of core radius and corresponding total radius; and
(3) energy of compression of the rock-substance.

A good approximation to the gravitational energy of a planetary body[†] with only a mantle and core is

$$W = W_0\{1+(\tfrac{4}{5}\zeta - 1)p^{4}\},$$

where
$$W_0 = \tfrac{4}{3}\pi^2 Ga^5\rho\rho_{\mathrm{m}}.$$

[†] The expression is obtained from $dW = rg\,dm$ with the assumption that $g = g_0$, a constant throughout the mantle, and $g = pg_0$ in the core.

$p = r/a$, the relative core radius; $\xi = \rho_c/\rho_m$, the ratio of densities of core and mantle material. Per unit mass the energy represented by W_0 is $\pi G a^2 \rho_m = 3 \times 10^4 \, \text{kJ kg}^{-1}$. Of this a proportion $(\frac{4}{5}\xi - 1) \approx 0.2$ will be released as the core diminishes from $p = 1$ to $p = 0$. Thus about $6700 \, \text{kJ kg}^{-1}$ will be released, an amount equivalent to about twice the initial thermal energy.

Note especially, however, that the rate of release of this energy will fall off rapidly, in fact proportional to $r^3 \, dr/dt$ with $r \to 0$. This source is important early in the geological life of a planet but, as we shall see, for the earth its contribution is now negligible.

Energy of compression

Because of the compressibility of rock-substance a planetary body is rather like a rubber ball. Work is needed to squash it, and if at a later time we let go it can do work. Here the ball is squashed by its own internal gravitational forces. Consider unit mass, of volume $v = 1/\rho$, compressed under a pressure P. The extra work required to compress the volume a further amount dv is $dU = -P \, dv = P \, d\rho/\rho^2$. Thus, given an equation of state $P(\rho)$ relating P and ρ we can integrate and obtain the total work required to compress the material from an initial density ρ_0 at zero pressure. For the equation of state used in this book:[†]

$$U = \{(\rho/\rho_0)^2 + 2\rho_0/\rho - 3\}\rho_0^2/6C.$$

Using typical values for the earth we obtain $W \approx 4 \times 10^3 \, \text{kJ kg}^{-1}$.

This is quite a lot of energy but only a portion of it is available. In the model of a freezing earth, the average values obtained by integrating through the volume of the earth give: completely liquid, $4000 \, \text{kJ kg}^{-1}$, completely solid, $1800 \, \text{kJ kg}^{-1}$. This total possible change of $2200 \, \text{kJ kg}^{-1}$ is comparable with that of the initial thermal energy.

Thermal energy

For an initial mean temperature of $3000\,^\circ\text{C}$ of material of specific heat $1 \, \text{kJ kg}^{-1} \text{K}^{-1}$ the internal energy is $3000 \, \text{kJ kg}^{-1}$.

There is a further source of internal energy available from that given up as material freezes. In this book all this energy is assumed to be released at the core–mantle boundary. Consider the freezing of unit volume. The energy change is

$$\rho_m \mathcal{H} \equiv \rho_c(C_c \theta + L) - \rho_m C_m \theta,$$

where ρ_c is the density of the core liquid, ρ_m the density of the mantle solid, C_c and C_m are the specific heats of the core and mantle material respectively;

[†] Putting $n = 3$ in $U = \{(\rho/\rho_0)^{n-1} + (n-1)(\rho_0/\rho) - n\}\rho_0^{n-1}/n(n-1)C$, where $C = \chi_0 \rho_0^n$.

L is the latent heat of fusion, θ is the freezing temperature, and $\mathscr{H}$ is the energy release per unit mass. Taking $C_c = C_m = 1\,\text{kJ}\,\text{kg}^{-1}\,\text{K}^{-1}$, $\theta = 2500\,^{\circ}\text{C}$, $L = 300\,\text{kJ}\,\text{kg}^{-1}$, and (from Chapter 6) $\rho_c/\rho_m = 1\cdot5$, we have $\mathscr{H} = 1700\,\text{kJ}$ kg^{-1}. This estimate of the enthalpy change on freezing of the core material will subsequently be assumed to be independent of temperature.

Radioactive sources

The radioactive decay, notably of the elements uranium, thorium, and potassium, provides a heat source which diminishes with time.

Let us, for example, assume that 10 per cent of current heat flow arises from radioactive decay, the bulk of the contribution arises from ^{40}K with decay constant $\lambda = 5\cdot4 \times 10^{-10}\,\text{yr}^{-1}$, and the amount has diminished to 10 per cent. Then, noting that if the rate of production is $p = p_0\exp(-\lambda t)$, the total is p_0/λ, we find the total energy available at the start to be $1\cdot5 \times 10^{30}\,\text{J}$. This is about $250\,\text{kJ}\,\text{kg}^{-1}$ with an initial rate of production $4\cdot6 \times 10^{-12}\,\text{W}\,\text{kg}^{-1}$.

It is important to emphasize that in our convective model the location of this material does not matter. Very probably it is concentrated in the upper mantle and crust, but for this model it could be distributed thoughout the volume of the earth.

TABLE 5.2
Radiogenic heat

	^{40}K[a]	^{232}Th	^{238}U[b]
Decay constant λ ($10^{-10}\,\text{yr}^{-1}$)	5·4	0·50	1·54
Power ($10^{-4}\,\text{W}\,\text{kg}^{-1}$)	0·29	0·26	0·94
Concentration (p.p.m.)			
Dunite[c]	0·1	0·01	0·04
Oceanic basalt	0·26	0·18	0·11
Olivine–basalt	1·30	3·50	0·90
Relative proportions now, as dunite	0·65	0·28	0·07
Relative proportions at 5×10^9 years B.P.[d]	0·95	0·014	0·033

[a] $^{40}\text{K} = 1\cdot2 \times 10^{-4}$ total K.
[b] $^{235}\text{U} \approx 0\cdot007\,^{238}\text{U}$; ignored.
[c] A uniform earth of dunite: equivalent heat flux now $56\,\text{mW}\,\text{m}^{-2}$.
[d] Total equivalent heat flow at 5×10^9 years B.P. higher by a factor of $9\cdot5$.

Energy of motion

The earth–moon system itself provides a source of energy through the relative arrangement and movement of its parts:

(1) the mutual gravitational energy of the system, $GmM/r \approx 7\cdot7 \times 10^{29}\,\text{J}$ or about $120\,\text{kJ}\,\text{kg}^{-1}$.

(2) the rotational energy of the daily rotation of the earth about its axis, $\frac{1}{2}I\omega^2 \approx 2\cdot1 \times 10^{29}\,\text{J}$ or about $35\,\text{kJ}\,\text{kg}^{-1}$.

(3) the energy of the orbital motion of the system is largely the kinetic energy of the moon's motion relative to the earth, $\frac{1}{2}v^2 m_{moon}$ $\approx 3.9 \times 10^{28}$ J, or about $7\,\mathrm{kJ\,kg^{-1}}$.

Owing to tidal interaction between the moon and the earth, these quantities have undoubtedly changed greatly through geological time and have profoundly affected the orbit of the smaller body, but as geological energy sources they are negligible.

TABLE 5.3
Energy stocks

Source	Energy $(\mathrm{kJ\,kg^{-1}})$
Gravitational	30 000
Compressional	4000
Thermal	3000
Freezing at core–mantle boundary	1700
Radiogenic	250
Mutual gravitational energy of earth–moon system	120
Rotation of earth	35
Orbital energy of earth–moon system	7

NOTES
1. Units are $\mathrm{kJ\,kg^{-1}}$ obtained from (total energy at 5×10^9 years B.P.)/(mass of earth).
2. $1\,\mathrm{kJ\,kg^{-1}} \equiv 6 \times 10^{24}\,\mathrm{kJ}$.
3. A supply of $1370\,\mathrm{kJ\,kg^{-1}}$ dissipated to exhaustion at a uniform rate in 10^{10} years will maintain a surface power flux of $50\,\mathrm{mW\,m^{-2}}$.
4. The solar flux into the earth's surface at the present rate for 5×10^9 years is equivalent to $5 \times 10^6\,\mathrm{kJ\,kg^{-1}}$. But nearly all this is re-radiated.
5. Our food has 4–$20\,\mathrm{MJ\,kg^{-1}}$. A human uses as food about $2 \times 10^{11}\,\mathrm{J}$ in a lifetime.

6. Building the reactor

NOW THAT all the ingredients are ready let us put the earth together. We commence our task at about 5000 million years B.P.

Has the interior always been hot?

Of course, I don't really know if the interior has always been hot. But I can assure you that it may well have been so. For this we thank Mr. Laplace. Let us run time back by blowing up the earth until it becomes a cloud in space as it might have been about 5000 million years ago, just a patch in a much bigger cloud destined to become our solar system. Suppose the gravitational energy in our little cloud is converted into heat as the cloud collapses under its own mutual gravitational attraction. It will get hot and some of the energy, say a proportion $(1 - \xi)$ will be lost as radiation. By the time the entire earth has condensed to roughly its present radius the temperature will be $T \sim \frac{4}{3}\pi \xi G \rho a^2/c$. To cut a long story short, for $\xi = 0.08$ we have $T \sim 3000\,\mathrm{K}$. If you live in a modern urban environment you won't need much persuading to believe that a relatively small proportion of matter in the atmosphere traps a lot of radiation. Equally, even when the radius of the cloud has shrunk to $10a$, so that 90 per cent of the gravitational energy is still to be converted to heat, the density of the cloud is about $5 \times 10^{-3}\,\mathrm{g\,cm^{-3}}$: about 4 times that of our present atmosphere. The transparency will thus be greatly reduced. So the earth could well have been hot at the time of its formation, and in this book I stay with Mr. Laplace.

We can envisage the final stages of this process in which the heat loss is more rapid than the release of gravitational energy as occurring in three fairly distinct phases.

1. *Gas phase.* The hot gas loses energy by radiative transfer and slowly collapses until a fog of liquid drops forms and in the outer parts some dust.

2. *Liquid phase.* The liquid globe will gradually form, but early in this phase it will be like a liquid foam. The surface temperature will be high and the rate of loss of energy from the interior will be controlled by radiation from the surface.

3. *Skin phase.* After a period of the order of 10^1 years† the surface

---

† Obtained from the model of Chapter 9 using a viscosity appropriate to a magma and with the surface temperature determined by heat flux $f = \xi\sigma(T^4 - T_0^4)$, where σ is Stephan's constant, T_0 is the reference temperature $\sim 300\,\mathrm{K}$, and ξ is a measure of the transparency of the primeval atmosphere, taken to be ~ 0.5.

temperature falls sufficiently for the first patches of solid to form briefly on the surface before foundering in the more dense liquid.

The crustal slag and the beginning of geological time

In this view, the acidic parts of the crust were the first persistent parts of the solid earth. When originally formed, this material would accumulate in rafts of loose material floating about like scum on a pond of liquid rock-substance, most of it being remelted. But ultimately these rafts would become sufficiently contiguous for there to be a nearly continuous skin of solid rock-substance enveloping the globe. It would be fragile and suffer frequent and extensive fragmentation but the appearance of that skin would be the signal that geological time had commenced, for there is then the possibility of a permanent geological record. We don't know if there is any of the original record still preserved. In view of what has happened to the crust since, it is unlikely—in fact the oldest known rocks are metasediments and not igneous.

In this book that moment is taken as 5000 million years B.P.

Core-mantle density contrast

We envisage an original liquid earth being cooled at the surface and freezing downward. It is not yet half-frozen. For a solid surface zone to grow downwards it is necessary that the density of material in the surface zone be

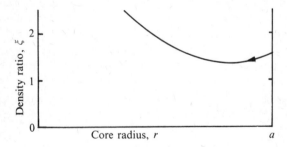

FIG. 6.1. Ratio of densities of top core liquid to bottom mantle solid, ξ, as function of core radius r; not adjusted to allow for compressibility. Note that $\xi > 1$: the core fluid has been more dense than the solid at the base of the mantle throughout freezing.

less than that in the liquid mass below. This is certainly the case now, since there is a large increase of density across the mantle–core interface.

By taking the known densities of the present mantle material we can as it were run the freezing backwards in time by progressively melting more and more of the lower mantle and thereby estimate the density contrast throughout the formation of the mantle. As we see, the density ratio, that is, the density of the core liquid divided by the density of the solid at the base of the mantle, has always been greater than unity. Indeed, it has always exceeded 1·3, and has had an average value of about 1·5.

Thus the mantle has grown and is continuing to grow downwards in the manner of ice forming on a pond.

Segregation of the substance in the earth

Clearly the substance in the core is different from that which forms the frozen base of the mantle. For otherwise, as we know, the liquid density would be about 10 per cent less than that of the solid. All known rock-substance behaves thus and *not* in the anomalous manner of ice.

From a given liquid rock-substance, as found in common experience of everyday chemical processes and the more detailed studies of experimental petrology, a variety of solid end-products is possible: this is the geological notion of differentiation. Thus, from the original liquid, namely, that obtained by melting the entire earth and mixing well, we envisage a sequence $S = (\alpha, \beta, \gamma, \ldots)$ of 'components' $\alpha, \beta, \gamma \ldots$ which become 'permanent' solids in the order $\alpha, \beta, \gamma \ldots$, the component α being the first to appear as the body cools from its entirely liquid state. Certain immediate statements are possible.

1. Even if the original rock-substance were defined and fixed for all possible bodies the sequence S will be a function of the particular thermal history of the body. The component deposited at the mantle–core interface will depend notably on the pressure and temperature there and on the current state of the liquid, whose rock-substance is depleted by the components already deposited. Experimental petrology shows that there is great variety of possible components even with rather small variations of liquid composition, temperature, and pressure. Thus, even though we could expect the sequences to be broadly similar they will differ greatly in detail.

2. Assuming that the temperature increases monotonically with depth through the evolution of the body, that the pressure does (this is certainly true), and that the core material is rapidly circulated so that it is well mixed, we must have two conditions simultaneously satisfied. Broadly these are that both the density of the component and its melting temperature† must increase with the order of deposition. Thus, if the subscript i refers to the i-component, if ρ is the liquid density, and T the melting temperature: $\rho_i < \rho$ and $T_i < T_{i+1}$. If the density is otherwise, once a kilometre or so has been frozen on that material will fall off the base of the mantle back into the liquid. If the melting-temperature relation is otherwise, any such material which may locally be formed will subsequently be remelted.

   It is necessary therefore to envisage the core saying to itself 'What can

---

† Melting temperature of a component refers to the melting or freezing temperature of a possible component if it should appear as a solid in the ambient liquid.

I get rid of now?' and reorganizing its material to encourage low-density, low-melting-point combinations. Of course, these in fact happen effectively at random and only certain combinations can survive.

3. In so far as the system as a whole moves towards the state of minimum gravitational potential energy at the greatest possible rate, we would have also $\rho_i < \rho_{i+1}$. I expect this to be generally the case, but it will occasionally turn out that it is incompatible with the other mandatory condition $T_i < T_{i+1}$.

4. The sequence S merely refers to the order of deposition from the liquid and does not imply where the components are deposited. The above discussion has concentrated attention on the mantle–core interface, but where components are both below their melting temperature and denser than the liquid they will accumulate in the central portion of the core.

In the present 'layering' in the earth (crust $2 \cdot 7 \, \mathrm{g \, cm^{-3}}$ and $3 \, \mathrm{g \, cm^{-3}}$, upper mantle $3 \cdot 3 \, \mathrm{g \, cm^{-3}}$, lower mantle $4 \cdot 1 \, \mathrm{g \, cm^{-3}}$, core $6 \cdot 2 \, \mathrm{g \, cm^{-3}}$ at surface temperature and pressure), the components are certainly more dense with order. In addition they are more basic. This is certainly the case for the crust and upper mantle, indicating an increase with order of the proportion of heavy atoms such as iron. Thus, as the lighter, lower-melting-point components are frozen out the concentration of heavy atoms in the core liquid increases.

I do not wish to imply that a 'component' is necessarily chemically homogeneous throughout its layer. Rather I have in mind that a closely related group of minerals will be present: for example, the olivines with their continuous solid-solution series.

Theoretical sketch: rate of core formation

Let the heat flux through the earth's surface be $f(t)$ with an average $\bar{f}$ over geological time. Assume that the heat flux across the core–mantle interface f_c is such that the loss of internal energy is proportional to the volume. Hence $f_c/f = r/a$. Now the change in internal energy in freezing a further portion dr to the base of the mantle is $\mathscr{H}$, the enthalpy change on freezing. As the freezing proceeds we further assume that all the heat loss is derived from that previously stored in the material of the new frozen layer. Thus, $\rho \mathscr{H} \, dr = -fr \, dt/a$ and thence $r/a = \exp(-t/\tau)$, where $\tau = \rho a \mathscr{H}/\bar{f}$. Taking $\rho = 5 \, \mathrm{g \, cm^{-3}}$, $\mathscr{H} = 1700 \, \mathrm{kJ \, kg^{-1}}$, and the present $r/a = 0 \cdot 54$ requires $\bar{f} = 210 \, \mathrm{mW \, m^{-2}}$.

The actual value for $\bar{f}$ shouldn't be taken too seriously. Our sketch is very crude, but taking it at face value we can make two deductions:

(1) the mantle–core boundary has been growing downward throughout geological time at an exponentially decreasing rate and continues to do so;

(2) the earth has been losing heat throughout geological time at an average rate greater than that of the present day.

Theoretical sketch: concentration of lithophilic elements and others

Suppose the core radius has diminished to r and that the concentration of a constituent is C per unit mass. The total mass of the constituent in the core of volume V is then $\rho V C$. After a change of core radius dr, the change in the total amount of the constituent in the core is $d(\rho V C)$ and this change is balanced by the amount deposited in the new frozen layer of area A, an amount $\rho_s A \beta C \, dr$. Here ρ_s is density of new frozen solid and β the relative proportion of the constituent in the solid compared with that of the liquid. Hence, equating the two changes in amount of constituent, since the total is conserved, $C = C_0 (a/r)^\xi$, where C_0 is the original concentration in the liquid, $\xi = 3(1 - \rho_s \beta/\rho)$, and we have rather crudely taken ρ as a constant throughout the freezing process. The corresponding concentration in the solid is βC.

1. For large atoms there is a pronounced tendency to concentrate in the more acid rocks. This is the case of $\beta \gg 1$, and such constituents are called *lithophilic*. Then $\xi \ll 0$ and there is a rapid fall of concentration with depth. For example, with $\beta = 15$, $\rho_s/\rho = 1 \cdot 5$ we find that the concentration has fallen to half at a depth of 64 km.

2. On the other hand, there is a tendency for some atoms to avoid the acid rocks. Then $\beta \ll 1$ and $\xi \gg 0$ and there is an increasing concentration with depth. Thus, for $\beta = 0 \cdot 63$ the concentration at $r/a = 0 \cdot 5$ is still $0 \cdot 9$ of the original; the great bulk of this constituent remains in the core.

Detailed calculation: freezing model

To investigate the mechanical consequences of the freezing model in any detail it is necessary to perform the calculations numerically. This is done very simply by integrating a finite-difference representation of the equations:

(1) $dm/dr = 4\pi\rho r^2$ mass enclosed within radius r,

(2) $g = Gm/r^2$ gravity,

(3) $d\rho/dr = -Cg\rho^{-(n-2)}$ from $dP/dr = -g\rho$ and equation of state,

(4) $dI/dm = \frac{2}{3}r^2$ moment of inertia,

(5) $dW/dm = -rg$ gravitational energy,

(6) $dV/dm = U$ compressional energy, for U of p. 38.

We know that $m = M$, the total mass at $r = a$; $m = 0$ at $r = 0$; and $I = 0 \cdot 3306 M a^2$ for the present earth with $a = 6371$ km and core radius 3471 km.

The calculations used: $n = 3$; for the mantle $\rho_0 = 3 \cdot 55 \, \mathrm{g \, cm^{-3}}$, $C = 28 \, \mathrm{Mbar^{-1}}$; for the core $\rho_0 = 6 \cdot 8 \, \mathrm{g \, cm^{-3}}$; $C = 168 \, \mathrm{Mbar^{-1}}$.

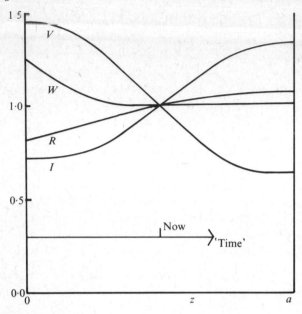

FIG. 6.2. Evolution of global structure: radius R, moment of inertia I, gravitational energy W, and compressional energy V, as functions of mantle depth z. Note that these data are derived on the assumption of a steady state, but in the text z is taken to increase with time. All values are relative to those of the present earth.

   The essence of the calculation is to integrate (1)–(3) repeatedly, each time adjusting the core radius until the mass is correct.
   We deduce several interesting consequences of the model.

1. The earth is expanding. The total possible change in radius of about 25 per cent has been going on nearly uniformly until now but most of the expansion has already taken place. The expansion in this model is dominated by the built-in feature of a permanent positive density contrast across the mantle–core interface. This relatively small expansion must be distinguished from that which would arise from cosmological changes in G—a quite unnecessary speculation.
     Note that in this model we make no statement about the rate of development; this is a study solely of possible arrangement of the parts.

2. The gravitational energy changes rapidly while the core is large, but the amount to be released for working the global machinery from now on is very small.

3. The moment of inertia increases throughout the freezing with its greatest rate of change about now: $\Delta I/I = 0.012$ in 100 million years. This will have a substantial effect on the rate of rotation of our globe.

There is, however, a somewhat larger and opposite effect on the rate of rotation due to the tidal interaction between the earth and the moon. The net effect has been to slow down the rotation rate.

Evolution after freezing

As the mantle forms it grows as a stratified body with density increasing downward. But this is by no means the end of the story.

1. As the over-all temperature changes as the body cools down and, to a much lesser degree, as the pressures adjust to the progressive over-all

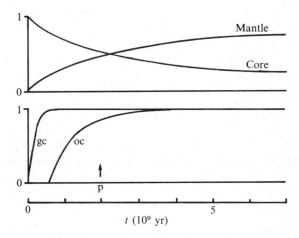

FIG. 6.3. Schematic development of the crust. Mass proportions of the granitic crust (gc) and the oceanic crust (oc) as a function of time t. The approximate time of the emergence of persistent land is shown as point p. The mass proportions of the mantle and core are shown for comparison.

mass distribution, the stratification will remain but different constituents will arise. This could be thought of as a form of global metamorphism. Occasionally rather more abrupt distinct phase changes will occur.

2. Superimposed upon these progressive unidirectional alterations there will be those arising from the vigorous eddying motion of the 'solid' mantle with its time-scale of the order of 100 million years. This mixing process will be at a maximum in the upper mantle.

3. Finally the rapid crustal rearrangements notably with the regurgitation of basaltic material result in further modification of the over-all stratification.

These simple considerations also allow us to make some fairly quantitative statements about the areal distribution of the crust. Suppose we make the following assumptions.

1. The interior was largely degassed by the beginning of geological time. The earth would be inundated to an average depth of about 2·5 km.

2. Granitic material will leave the mantle as quickly as possible.

3. The thickness of the granitic crust is determined by the criteria: isostasy, a bottom melting temperature of 700 °C, and rapid erosion to near sea-level for any crust above sea-level. Note that the total amount of granitic material corresponds to that in an upper crust of, say, 20 km thickness, over 30 per cent of the surface, a global average thickness of 6 km.

4. Lumps of granitic crust, once joined together, tend to stay together owing to the very much larger viscosity of granitic rock-substance than any other.

   Thus we have the scenario for the over-all development of the crust:

   (a) While the 700 °C isotherm is shallower than 6 km, there will be a more-or-less global skin of granitic rock-substance. For an interval of about 500 million years there will be no oceanic crust.

   (b) After the 700 °C isotherm is below 6 km all the granitic rock-substance has left the mantle. As the 700 °C isotherm deepens, 'spaces' begin to form between patches of granitic crust. We can now have granitic patches thicker than 6 km, the granitic patches reach down to the 700 °C isotherm, and there is no longer enough granitic rock-substance to cover the globe. Oceanic crust begins to form. The whole lot is still under water.

   (c) As the isotherm descends, the granitic crust continues to thicken and more oceanic crust appears. Local piles of granitic material may temporarily reach above sea-level, but ultimately the average level of the granitic crust will reach sea-level. What might be called the phase of persistent land occurs. For 20 km of granitic crust of bottom temperature 700 °C, the corresponding heat flow is about 100 mW m^{-2}, the value perhaps after 2000 million years. (A detailed study of the earth's thermal history is given in Chapter 9.)

   (d) Finally, the 700 °C isotherm descends below the level determined by a granitic crust in isostatic equilibrium and eroded to near sea-level. The areal distribution of the crust is no longer directly determined by the temperature at the bottom of the crust.

7. Dynamical system

THE EARTH is now constructed and the major structures—crust, mantle, core—are built in. But as planetary engineers our project would be a failure if we left it at that. Very little would happen: our planet would be still-born. We have yet to make it work by providing some vigorous mass- and energy-transfer process.

The spatial and temporal variability in the geological record requires vigorous dynamical processes within the earth. When we look around for natural mixing systems that operate on a large scale we find only one: turbulence. The turbulence familiar to all of us in the wind and streams clearly is able to mix up things vigorously, as we can see by putting out smoke or dye. This kind of momentum-turbulence is not what we want, since within the earth the flows are viscous. But there is another kind of turbulence—thermal-turbulence—which is found in large-scale vigorously convecting systems; this type of turbulence can occur in a viscous material. In the language of fluid mechanics momentum-turbulence has a high Reynolds number, thermal-turbulence a high Peclet number.

We start with a basic model of a homogeneous working material imprinted on which are solely variations of temperature. It is obvious, however, from the variety of rocks extruded from the earth's interior that the earth's mantle is by no means homogeneous. Nevertheless, it is not so much variations of chemistry but variations of physical properties that concern us first. There will undoubtedly be variations of all the physical parameters relevant to a dynamical system, but of these the enormous variations of viscosity will be dominant. The kinematic viscosity decreases very rapidly with temperature and increases strongly with the proportion of silica (SiO_2). At this time it is impracticable to cope with all the possibilities in a single model. We can, for example, make a model with a thin very viscous upper layer and a less viscous deep layer, thereby representing the viscosity variation in the earth by a step function. This approach can be extended by using a material which does have a reasonably large variation of viscosity such as petroleum jelly, or table jelly.

The temperature variation of viscosity does, however, provide one substantial advantage. The basic model shows us that over a given horizontal plane there will be temperature variations of the order of 10 per cent of the deep-mantle temperature. The consequent variations of buoyancy are, of course, what all this work is about, but a region of high temperature will be especially mobile because its viscosity is very much lower than that of its immediate surroundings. This situation can be simulated in the laboratory with a homogeneous viscous fluid simply by differential heating. So it should not be assumed that because in some models we have direct differential heating at the base of a layer of fluid representing a layer of the mantle we are assuming that this is actually what is happening.

After a rather detailed discussion of the basic model we will consider a number of modifications to it which arise from a crude attempt to model more accurately the viscosity variation. The key feature of these models is the interaction between the various parts of the total system. For example, a slab does not sit passively on top of the mantle but modifies the mantle structure near it, which in turn affects the behaviour of the slab. When I use the words 'slab' or 'sheet' I mean a vertically thin region: a portion of the more viscous crust and upper mantle that for a time moves with nearly uniform horizontal velocity either because it is much more viscous than its surroundings or because the horizontal stress distribution beneath it is rather uniform.

Heat and mass transfer in a fluid: general considerations

Motion through a fluid

Let us take a number of solid objects all of the same shape, differing only in a characteristic length scale h, held at a fixed orientation and totally immersed in a steady stream of kinematic viscosity v flowing with a uniform velocity u remote from the object. Each body will experience a force or drag due to the altered pressure distribution in the stream and to the viscous stresses on its surface. The flow about the body is specified by the three quantities u, v, h, which define a single dimensionless number $(Re) = uh/v$, the so-called Reynolds number. Thus any quantity of interest, expressed in dimensionless form, is solely a function of (Re). For the drag F we can choose a unit of force $\rho h^2 u^2$, where ρ is the density of the fluid. Hence the drag ratio $C \equiv F/\rho h^2 u^2 = f(Re)$. Data from numerous experiments confirm this statement.

Three distinct regimes can be distinguished.

1. Small Reynolds number, $(Re) \lesssim 10$. Here the flow is steady, the streamlines smoothly curve over each other, and $C \propto 1/(Re)$. This is called viscous or Stokesian flow.

2. Intermediate, $(Re) \sim 10^2$. This depends on the shape of the body, but in general for bluff bodies there is a pulsatory character to the flow. We are all familiar with this through the singing of telephone wires, a resonance between the mechanical properties of the strung wire and the pulsatory motion near the wire—the so-called aeolian tones, after an ancient wind-driven harp.

3. Large Reynolds number, $(Re) \gtrsim 10^3$. The flow near the body and for a long way downstream behind it is not at all steady. The wake region has the appearance of a rather chaotic sequence of short-lived eddying motions. We find $C \sim$ (constant). This is turbulent flow.

Which of these regimes applies to the earth? Consider a rather extreme case: a slab of 10^3 km extent moving at $3 \, \mathrm{m \, yr^{-1}}$ in a mantle of kinematic viscosity $10^{16} \, \mathrm{m^2 \, s^{-1}}$ for which $(Re) = 10^{-17}$. This is an exceedingly small Reynolds number. Undoubtedly such motions inside the earth are Stokesian! Only in the case of the extremely rapid flows found in volcanic systems or where actual fracturing occurs shall we need to consider other possibilities.

In the Stokesian régime, $F \propto \mu h u$. In other words, the forces on objects within the earth which arise from the action of flow are proportional to viscosity, size, and velocity. A good approximation for the earth is $(Re) \to 0$.

Forced convection

Suppose now that the object is being heated or cooled. For the moment take an extreme case in which the interior of the body is maintained at a uniform temperature, say by means of a vigorous mixing process inside. The heat transfer will then be controlled by the surrounding fluid, in particular by its thermal diffusivity κ. Thus our system is now specified by four quantities: u, v, h, κ, which define two independent dimensionless numbers. Since we have already seen that $(Re) \to 0$, rather than v, it is more relevant to choose κ as the unit of diffusivity. A suitable choice of dimensionless numbers is: $(Pe) = uh/\kappa$, the so-called Peclet number, and $(Pr) = v/\kappa$, the so-called Prandtl number.

What about the earth? Taking $\kappa = 10^{-6} \, \mathrm{m^2 \, s^{-1}}$, $v = 10^{16} \, \mathrm{m^2 \, s^{-1}}$, the Prandtl number $(Pr) = 10^{22}$, virtually infinite. Thus the statement $(Pr) \to \infty$ is a good approximation for the earth, and in forced convection the Peclet number is the sole parameter of importance. For the example above we have $(Pe) \sim 10^{-17} \times 10^{22} \sim 10^5$, indicating powerful effects of the flow on the heat transfer. Even with velocities of the order of $0.03 \, \mathrm{m \, yr^{-1}}$, $(Pe) \sim 10^3$.

Simple description of the state $(Re) \sim 0$, (Pe) large

When the Reynolds number is small, the effects of the inertia of the fluid are negligible and the flow is dominated by the effect of the viscous forces. These forces are felt throughout the flow volume but are most intense near the surface of the body. They are quite small beyond a distance h from the body. We can think of the situation thus. In a region with a velocity gradient fluid elements are being rotated. The rate of rotation, proportional to the so-called vorticity, is greatest near the surface of the body, where the fluid is brought to rest. Thus the flow is composed of the uniform incident stream, superimposed on which is a source of vorticity distributed over the surface of the immersed body. The vorticity diffuses outward with diffusivity v, very quickly when v is large, so that the vorticity distribution is always close to being in a steady state. Hence from a viscous stress $\sim \mu u/h$ acting on the area $\sim h^2$, the total force $F \sim \mu u h$. The ratio $F/\mu u h$ will clearly depend solely on the shape of the body; a sphere, for example, has the ratio 6π.

For a heated body in an otherwise isothermal incident flow the heat can diffuse a distance $\delta \sim (\kappa t)^{\frac{1}{2}}$, where κ is the thermal diffusivity of the fluid and

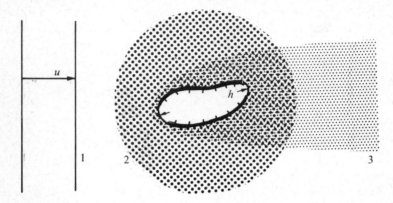

FIG. 7.1. Diagram for forced convection at large Peclet number. A cold uniform flow incident on a hot bluff body of typical dimension h. (1) Incident flow; (2) zone of viscous influence, of radius $\propto h$; (3) zone of thermal influence, of width proportional to $\sqrt{(\kappa x/u)}$, where x is distance downstream from the nose of the body.

t is the time a fluid element spends in the vicinity of the body. Since $t \sim h/u$, if κ is small only a restricted zone of fluid near the body is warmed. This is in complete contrast to the situation for the diffusion of vorticity where v is very large. Hence the heat flux $\sim K\Delta T/\delta$ acting over an area $\sim h^2$ produces a total heat flow from the body $\sim K\Delta T a^2 (u/ka)^{\frac{1}{2}}$. This is more simply expressed as the so-called Nusselt number (Nu), a dimensionless measure of the ratio of convective heat transfer to that which would occur by conduction if there was no convection, for which $(Nu) = h/\delta \sim (uh/\kappa)^{\frac{1}{2}} = (Pe)^{\frac{1}{2}}$. In other words the heat transfer depends solely on the Peclet number where the Prandtl number is large.

Free convection

Finally consider an enclosed body of fluid across which heat is being transferred by fluid motion because of temperature variations imposed over the surface of the body, but in which there is no imposed fluid motion. The motions will arise because of the differences in weight of adjacent columns of fluid, which are due to density differences produced by variations of temperature throughout the fluid body. Consider the very simple case in which the density ρ is related to the temperature T by $\rho = \rho_0\{1-\gamma(T-T_0)\}$, where $\rho = \rho_0$ when $T = T_0$ and γ is the so-called coefficient of cubical expansion. Thus, if the temperature over the surface of the body ranges between $T = T_0$ and $T = T_0 + \Delta T$ the scale of weight differences is $\gamma g h \Delta T$, where g is the local acceleration due to gravity. This system is then specified by the four quantities v, h, κ (as before but with u unspecified), and $\gamma g h \Delta T$, which define two dimensionless numbers: $(Pr) = v/\kappa$, the Prandtl number (as

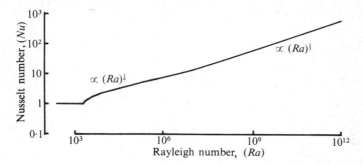

Fig. 7.2. Heat-transfer relationship for free convection in a wide uniform layer of a viscous fluid. The dimensionless heat flux (Nu) against Rayleigh number (Ra). Note that $(Nu) = 1$ corresponds to no convection. Note the convective cut-off below $(Ra) \approx 10^3$. Data from steady-state laboratory measurements with cooling from above *and* heating from below.

before), and $(Re) = \gamma g \Delta T h^3/\kappa v$, the so-called Rayleigh number. Clearly the Rayleigh number is a measure of the effectiveness of the buoyancy forces acting against the combined resistance to motion of viscosity and the diffusion of temperature variations by thermal conductivity.

For a body of given shape and given boundary temperature-distribution, the heat transferred across it measured by the dimensionless heat-transfer quantity (Nu) reveals three main regimes.

1. Low Rayleigh number, $(Ra) \lesssim 10^3$. Convective motions are extremely weak and $(Nu) \sim 1$. In other words the heat is transferred largely by ordinary thermal conduction.

2. Intermediate Rayleigh numbers, $(Ra) \sim 10^4$. Steady convective motions, often cellular with cells extending from one side of the cavity to the other; as (Ra) increases the heat transfer approaches $(Nu) \sim (Ra)^{\frac{1}{4}}$.

3. High Rayleigh numbers, $(Ra) \gtrsim 10^5$. Vigorous non-steady convection dominated by an apparently chaotic eddying motion with eddies of a wide range of sizes and $(Nu) \propto (Ra)^{\frac{1}{3}}$.

   As with forced convection, the influence of (Pr) is negligible for $(Pr) \gg 1$.

We can make an estimate of the earth's Nusselt number $(Nu) = fa/K\Delta T$

from $f = 50\,\text{mW m}^{-2}$, $K - 2\,\text{W m}^{-1}\text{K}^{-1}$, and, say, $\Delta T - 2500\,\text{K}$ to obtain $(Nu) = 64$. This suggests a large Rayleigh number. Let us therefore make an estimate of the Rayleigh number of the earth using the parameters $\gamma = 10^{-5}\,\text{K}^{-1}$, $g = 10\,\text{m s}^{-2}$, $h = a = 6370\,\text{km}$, $\kappa = 10^{-6}\,\text{m}^2\,\text{s}^{-1}$, and a possible and arithmetically convenient $v = 6\cdot46 \times 10^{16}\,\text{m}^2\,\text{s}^{-1}\,\text{cm}^2\,\text{s}^{-1}$. Then the Rayleigh number $(Ra) = 10^9$. The values for g and a are quite accurate. Values for γ, ΔT, and κ are probably accurate within a factor of 2; for example, there is no evidence that ΔT would lie outside the range 1500–6000 °C. Both γ and κ may be estimated low. The dominant uncertainty is for v. The kinematic viscosity of the mantle may well lie in the range 10^{15}–$10^{18}\,\text{m}^2\,\text{s}^{-1}$. Nevertheless, as we shall see more clearly in Chapter 8, the relevant values of this and the other parameters are those found in the uppermost part of the mantle. Further, the model of Chapter 9 allows us to pin down the value of the kinematic viscosity within much narrower limits, to values compatible with those determined from the rate of isostatic readjustment. For the present, however, it is sufficient to note that the Rayleigh number is large. In summary:

$(Re) \rightarrow 0$,   slow viscous flow;

$(Pr) \rightarrow \infty$,   heat transfer independent of inertial effects;

$(Pe) \sim 10^3$,   Peclet number effects dominant in forced convection;

$(Ra) \sim 10^9$,   free convection, strong and turbulent.

The weak convective state $(Nu) \frown 1$

When one is asked to describe how convection occurs there is often the immediate thought, 'hot air rises'. Unfortunately this notion gives an erroneous emphasis. It is not enough to refer solely to the buoyant element; we need also to consider its surroundings. We shall then also observe at the same time not only that 'cold air descends' but that the 'cold air' is in a different place from the 'hot air'. Convection is maintained by *horizontal* variations of temperature. Consider an otherwise uniform layer in which a vertical column is heated. Let us compare the weights of two vertical columns of fluid of equal area: one the hot one, the other in the fluid at ambient temperature. Clearly the weight of the hot column is less than that of the ambient column, since it is less dense. The fluid pressure at the base of the ambient column is thus higher than that at the bottom of the hot column. A horizontal pressure gradient exists between the bases of the columns, and it will be of the order of $\gamma\rho_0\,g\theta h/l$ for a temperature difference θ, layer depth h, and distance between the columns l. Fluid will then begin to flow horizontally out from the base of the ambient column toward the base of the hot column. Since the total quantity of fluid must be conserved, the loss from the base of the ambient column will be made up by a corresponding flow into the top of the ambient column. Thus a circulatory flow is set up.

Now consider a uniform layer of fluid of large horizontal extent, across which is maintained a large uniform vertical temperature difference by maintaining the upper surface at a low temperature and the lower surface at a high temperature. Let us go to a lot of trouble to ensure that the two temperatures are held steady. There is no imposed horizontal temperature gradient. Yet we find the layer vigorously convecting! Indeed, if we measure the horizontal variation of temperature, say along a horizontal line in the fluid, we find large horizontal temperature gradients. How, in spite of all our precautions, has this come about? Plainly if the temperature field was such that over every horizontal plane there was no variation, there would be no motion, since there would be no net buoyancy forces to drive the motion. We are forced to the conclusion that our experimental apparatus is not as perfect as we imagined. Of course it is not: above the critical Rayleigh number the system is unstable, and any disturbance, no matter how weak, will grow at a rate that is initially exponential until it is of finite amplitude. This amplification of effectively random disturbances produces the horizontal temperature variation.

On the other hand, if the Rayleigh number is small enough, the disturbances are damped and persistent convection is not possible. The nature of the combined role of the two molecular processes of diffusion and viscosity in this process is best revealed in the classic stability problem of a thin layer of fluid across which is a uniform density gradient $\rho(z)$ such that more dense fluid is at the top. Motion does not necessarily ensue even though the system seems unstable.

Theoretical sketch: the combined stabilizing role of viscosity and conductivity

The stability of a system is usually determined by giving it a small nudge and observing its subsequent behaviour. If the system ultimately returns to its original state it is said to be stable; if not, it is unstable. We shall nudge our system by considering the motion of a portion of the fluid, a spherical volume of radius r, which is given a density excess ρ' over its ambient density. If the density excess decreases with time in all circumstances, the system is stable.

Although our fluid particle is indistinguishable, apart from its density excess, from the surrounding fluid we consider it as a separate entity.

The particle will experience a buoyancy force proportional to its density excess because it has a density different from its surroundings. It will begin to move vertically. But as it moves it experiences a viscous drag proportional to its velocity. Thus if at time t the vertical velocity is w (we measure z downwards here), the inertia, buoyancy, and viscous forces (per unit volume) together give the equation of vertical motion:

$$(\rho + \rho')\frac{\mathrm{d}w}{\mathrm{d}t} = g\rho' - \frac{\rho v w}{k},$$

where $k = Xr^2$ with $X - \frac{4}{9}$ for a sphere of radius r. In a porous medium k is equal to the permeability (see Chapter 13).

1. Provided the density excess is sufficiently small so that $\rho' \ll \rho$, the inertia term is effectively $\rho\, dw/dt$. In this case the contribution of the density excess to the motion is solely due to the buoyancy force $g\rho'$. This apparently innocuous simplification is in fact extremely far-reaching. It is known as the Boussinesq approximation.

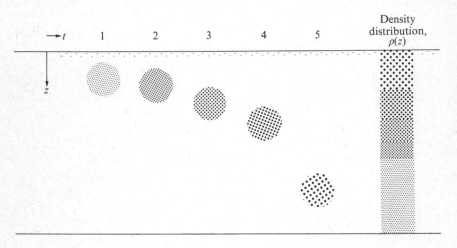

Fig. 7.3. Diagrammatic illustration for discussion of the stabilizing effect of both viscosity and thermal conductivity on a layer of fluid cooler, and thus denser, at the top than the bottom. Note the density distribution $\rho(z)$. A fluid parcel, its density contrast to its surroundings indicated by the shading density, is shown at various instants in time. The parcel actually falls vertically.

2. The particle will accelerate from rest, with an initial acceleration $g\rho'/\rho$, and speed up, but as it does so the viscous drag increases. After a finite time† of the order of $\tau = k/v$, the buoyancy and viscous forces will come into balance and the acceleration will fall to zero. The particle has reached its so-called limiting velocity, $W = kg\rho'/\rho v$ (obtained by setting $dw/dt = 0$). The distance travelled in time τ will be of the order of τW, and provided that ρ'/ρ is sufficiently small this distance will be very small compared to the width of the layer. Thus, because the acceleration time is relatively short, we have $dz/dt \approx W$ throughout most of the motion. The inertial effect of the fluid is negligible.

3. As the particle moves, its density excess changes owing to two effects: it finds itself in a new ambient density and yet density diffusion is continually

† The complementary function of $\{(d/dt)+(v/k)\}w = g\rho'$ is $w = w_0\exp(-t/\tau)$ with $\tau = k/v$.

trying to equalize its density to that of its surroundings. Thus, in a time-interval dt

$$d\rho' = d\rho - \rho' \, dt/T,$$

where dρ is the change due to the particle being in a new ambient density, and the density change due to diffusion is crudely represented by $-\rho' \, dt/T$, where the diffusion time-scale $T = r^2/\varepsilon\kappa$ with $\varepsilon \approx 10$ for a sphere of radius r and diffusivity κ.

4. Hence, taking $\rho' = \rho'_0 \exp(nt/T)$, so that when $n > 0$, the system is unstable, when $n < 0$ it is stable, and when $n = 0$ it is marginally stable, we find

$$n = \alpha p - 1,$$

where $\qquad \alpha = Xgr^4/\varepsilon\kappa v$ and $p = -(1/\rho)(\partial\rho/\partial z)$.

(a) For $p < 0 : n < 0$. A fluid layer heavier at the bottom is stable. No surprise!

(b) For $p > 0$.

(i) If either κ or $v \equiv 0$, $n = \infty$ and the system is unstable. Thus unless *both* molecular processes are present the stable state is impossible. This is what one would intuitively expect. Henceforth we take both κ and v to be finite.

(ii) With a particle of given radius we can always find a sufficiently small but non-zero negative density gradient for which $n < 0$ and the system is stable.

(iii) On the other hand, given the density gradient, the most unstable case is then for r as large as possible, namely, $r \approx h/2$. Marginal stability is then reached with $n = 0$ and $\alpha p - 1$, when

$$\frac{gh^4}{\kappa v}\left(\frac{1}{\rho}\frac{d\rho}{dz}\right) \approx \frac{16\varepsilon}{X} \approx 10^3.$$

But this is just our old friend the Rayleigh number, slightly disguised.

Thus we see the subtlety of the situation. The existence and nature of an overlying stable layer of more dense fluid arises from the interplay of the combined action of the damping of motion by viscosity and the ironing-out of density variations by diffusion, which removes the source of buoyancy. Neither of these processes alone is able to stabilize the layer.

It is worth noting that the density gradient does not have to be uniform in the situation described here. The case with a uniform gradient is merely a simple archetypal situation of great analytical interest. In practice a strictly linear gradient is never encountered.

Simple description of vigorous convection: the state $(Pr) \sim \infty$, (Ra) large

The observation that for a container of fluid vigorously convecting the heat transfer parameter (Nu) is proportional to $(Ra)^{\frac{1}{3}}$ is, among other things, a roundabout way of stating that the power passing through the container is independent of the depth of the container. This can be demonstrated directly. In a given experiment on vigorous convection we measure the power and the temperature difference. It is then a simple matter without otherwise altering

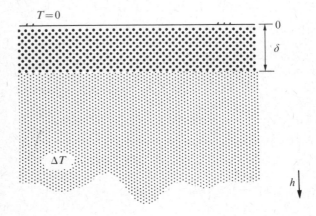

FIG. 7.4. Model of the temperature field for a deep fluid body cooled from above. The light stipple represents the interior region of uniform mean temperature. The heavy stipple represents the thin sublayer, attached to the upper surface, across which the temperature is regarded as varying linearly from the surface temperature, 0 to ΔT at the bottom of the sublayer.

the apparatus to pour in more working fluid and repeat the measurement. When the system is again in equilibrium, neither the measured power nor the temperature difference will have made a net change. Since the power transferred for a given temperature difference is independent of layer depth we deduce that the power transfer is controlled in some fixed region of the flow and that that region must be attached to the surface.

In a contained system the dynamical processes are profoundly modified near the container walls. In the earth the upper surface is externally constrained to be nearly at a fixed temperature and in addition is much stiffer than the interior. Thus there will be a region immediately below the upper surface where the damping effect of the molecular processes of thermal conductivity and viscosity will be important. This region, called the thermal sublayer will be considered in detail in the following chapter. Here let us just make the gross assumption that the molecular processes will be strictly confined to an upper layer of thickness δ above a uniform interior, and that this upper layer is in a marginally stable state. By that I simply mean that the vigorous interior flows continually tend to thin the sublayer until it can barely overturn. It must, however, be able to overturn, or else the convection will cease. Thus,

$$\frac{\gamma g \, \Delta T \, \delta^3}{\kappa \nu} = (Ra)_{\mathrm{c}} \sim 10^3,$$

where $(Ra)_{\mathrm{c}}$ is a Rayleigh number related to the sublayer thickness and with the entire temperature difference across the sublayer, corresponding to the condition for onset of overturning of the sublayer. Since the heat is transferred across a layer in a marginal state at the same rate as it would be by conduction, and since the Nusselt number $(Nu) = h/\delta$, we have $(Nu) = \{(Ra)/(Ra)_{\mathrm{c}}\}^{\frac{1}{3}} \approx 0\cdot1(Ra)^{\frac{1}{3}}$.

Thus, so far as the gross properties are concerned, the role of molecular processes is confined to a thin upper layer. These processes have a negligible role in the deep interior.

8. Cauldron

NOW LET US, for the moment, ignore all the obvious and complex details which we see all too clearly manifest in the earth's crust. Let us concentrate our attention on the sort of processes that go on in the great bulk of the earth's mass. Envisage a global, ubiquitous process pervading, as it were, a naked primitive earth. To get some idea of the possible processes we shall study a laboratory model, a miniaturized version of a hot earth cooled at its surface. Our model is simply a pot of hot oil, broader than it is deep, insulated on the base and sides, and cooled from above. Although I give a description appropriate to the laboratory you should mentally translate that into global terms. Table 8.1 (p. 64) is designed to help that translation. Of course, what I describe is actually a pot of hot oil cooling down, but since the results are for models which have the same dynamical proportions as a hypothetical viscous earth, in so far as the earth behaves viscously the laboratory results are merely for a smaller system in which the processes are greatly speeded up. They are actually speeded up by about 10^{14} times. This chapter is not so much about fluid dynamics as about the thermodynamics of a naked earth.

Thermal turbulence

What are the properties of a body convecting vigorously at a Rayleigh number of the order of 10^9? Fortunately Rayleigh numbers of this order are readily achieved in the laboratory, for example in a 30-cm-deep layer of medicinal paraffin at about 30 °C above room temperature. So now let us go into the laboratory and see what we can find out about this vigorous convection.

In everyday speech the meaning of 'thermal turbulence' is clear. We refer to the more-or-less chaotic motion produced by heating a large body of fluid. However, a more precise definition is difficult. There is no problem with the word 'thermal', although the word 'buoyancy' is possibly preferable. We consider motions generated by buoyancy forces, namely, forces which arise from spatial variations of an otherwise uniform imposed force field. It is simplest to consider the case in which these variations arise from spatial variations of density in a uniform acceleration field. A powerful simplification of the description of the flow field is possible with the so-called Boussinesq approximation in which density variations relative to the reference state are ignored except in so far as they generate buoyancy forces. This approximation may still be a good one even if in the reference state there are large relative variations in the density field. There are three effects of the

FIG. 8.1. Turbulent convection at $(Ra) = 2.5 \times 10^7$. Visualization of the flow. Photograph of the mid-vertical plane. Depth of fluid 6·0 cm, diameter of container 20·95 cm; 1 s exposure. This flow is not steady; the flow is continually changing. The photograph emphasizes the large eddies.

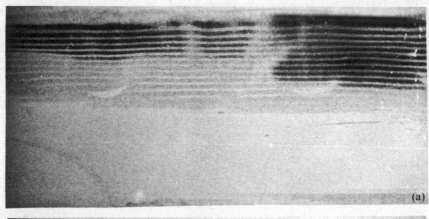

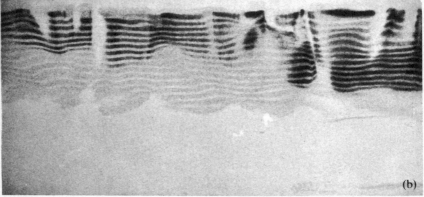

FIG. 8.2. Visualization of the temperature distribution; (a) initial appearance of blobs, shortly after surface cooling begins; (b) final statistically steady state. As in Fig. 8.1, this temperature pattern is not steady.

temperature field in this approximation: vorticity is generated by the gradients of the bouyancy forces normal to the applied force field; temperature is advected by the flow, and diffused by molecular processes.

On the other hand, the word 'turbulence' poses a difficult and contentious problem. At the outset we require a definition of turbulence and a specification of the circumstances under which it is present. While there can be little doubt that the turbulence generated in a shear flow, such as in a river, is fundamentally different form laminar or quasi-laminar flows, the distinction for turbulence generated by buoyancy forces is not so clear. Nevertheless, studies of a horizontal layer of fluid uniformly cooled from above do show a remarkable change in the flow at a Rayleigh number of the order of 10^6. Flows at higher Rayleigh numbers exhibit the property of 'continuous instability.'

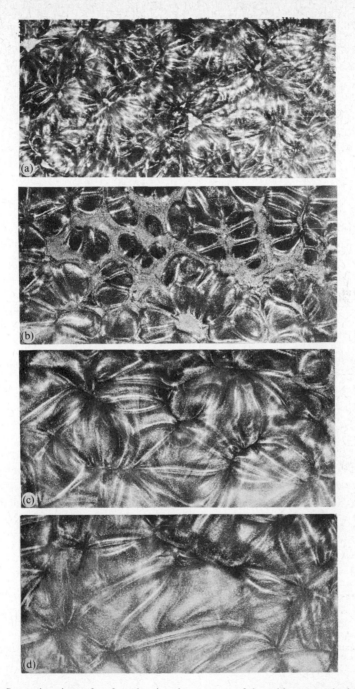

Fig. 8.3. Successive views of surface showing the structure of the sublayer as a body of fluid cools down. The photographs show an area 20 cm × 10 cm; container of depth 30 cm. Note that these patterns are continually and rapidly changing (in intervals of the order of 1–10 s), as compared with the gradual rate of cooling and growth of sublayer thickness (over intervals of the order of 10^3 s). The Rayleigh number range during the sequence (a)–(d) is from 10^9 to 10^6.

Methods of describing turbulence fields

Two different methods of describing the properties of the turbulent system
are used.

1. *Statistical approach.* For any variable of interest we represent its variation
   either in space or in time by a suitable average or mean quantity, together
   with its variation about the mean; and this variation itself will be described
   in terms of its average. Thus, for example, the total temperature
   $\mathcal{T} \equiv \langle \mathcal{T} \rangle + \theta$, where $\langle \mathcal{T} \rangle$ is the mean of $\mathcal{T}$ and θ is the so-called
   fluctuation about $\langle \mathcal{T} \rangle$. In a system which is statistically steady in time
   and statistically homogeneous over horizontal planes, averages taken over
   a horizontal plane at a fixed time or averages taken over time at a fixed
   place on that plane will be equal. This occurs when the system-boundary
   fluxes are steady or the time of cooling of the system as a whole is long
   compared with the time-scale of the turbulence. Thus, $\langle \mathcal{T} \rangle$ is solely a
   function of depth z and not of horizontal position. By definition the
   average temperature fluctuation $\langle \theta \rangle = 0$ but $\langle \theta^2 \rangle$, the mean square, is
   not zero. We write $\theta' = (\langle \theta^2 \rangle)^{\frac{1}{2}}$, the r.m.s. (root-mean-square) temperature
   fluctuation. Where an average over the whole flow space is required I
   shall write—for example, $\bar{\theta}'$ for the average of θ' over depth.

2. *Mechanistic approach.* We recognize distinct regions with distinct
   processes occurring in them and try as directly as possible to look at
   the individual fluid masses.

Where at all convenient, data are presented for experiments at $(Ra) = 10^9$,
large Prandtl number, and with temperatures, lengths, and velocities in the
units ΔT, h, κ/h, where ΔT is the horizontal mean temperature excess at
the base of the apparatus. Some of the data were obtained under somewhat
different circumstances but have been scaled accordingly.

TABLE 8.1
Properties of thermal turbulence of a cooling layer
Comparative values for an earth at $(Ra) = 10^9$, $\Delta T = 2500\,\mathrm{K}$

Minimum Rayleigh number	10^6	
Surface heat flux, mean (Nu)	$(Nu) \approx 0{\cdot}1(Ra)^{\frac{1}{3}}$	~ 100
Sublayer thickness δ	$\gamma g\, \Delta T\, \delta^3/\kappa\nu \approx 10^3$	$\sim 10^2\,\mathrm{km}$
R.m.s. temperature fluctuation $\bar{\theta}'$	$\bar{\theta}'/\Delta T \approx 2(Ra)^{-\frac{1}{3}}$	$0{\cdot}063: \sim 160\,\mathrm{K}$
R.m.s. velocity $\tilde{q}$	$h\tilde{q}/\kappa \approx 0{\cdot}3(Ra)^{\frac{1}{2}}$	$0{\cdot}05\,\mathrm{m\,yr^{-1}}$
Mean velocity $\langle \mathbf{q} \rangle$	0	0
Global temperature $\bar{T}$	$\sim \Delta T$	$2500\,\mathrm{K}$
Spectral cut-off frequency	κ/δ^2	$\sim 10^{-8}\,\mathrm{cycle\,yr^{-1}}$

Average properties

The distribution of mean temperature, written $T = T(z) \equiv \langle \mathcal{T} \rangle$, shows two distinct regions.

The sublayer

1. A region near the upper cold surface, of a large vertical temperature gradient of extent δ such that $\gamma g \, \Delta T \, \delta^3 / \kappa v \approx 10^3$. The temperature gradient is quite uniform immediately below the surface.

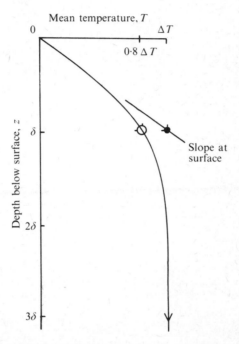

FIG. 8.4. Vertical profile of horizontal-mean temperature at Rayleigh number $(Ra) = 10^9$. Detail defining δ, the sublayer thickness. The mean temperature continues at its maximum value right to the base of the flow.

The interior

2. A region below the sublayer, of uniform mean temperature. The bulk of the fluid is isothermal in the mean. In a large system, as opposed to the small laboratory model, this region will be isentropic in the mean. The mean velocity $\langle \mathbf{q} \rangle$ is zero.

The fluctuating part

These are the temperature fluctuation θ and the velocity $\mathbf{q} = (u, v, w)$. Because the fields are horizontally homogeneous the statistical properties of u and v are identical; of these two, we refer separately only to u.

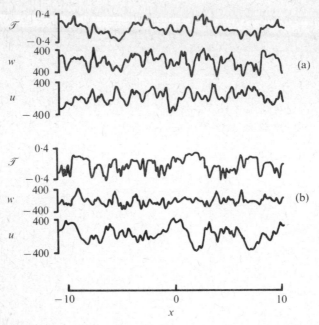

FIG. 8.5. Spatial variation at one time of temperature $\mathcal{T}$ relative to the local mean temperature, vertical velocity w, and horizontal velocity u, at $(Ra) = 10^9$. (a) Deep interior; (b) base of sublayer. Length unit, container depth h, velocity unit κ/h, temperature unit ΔT. (Adapted from Deardorff.)

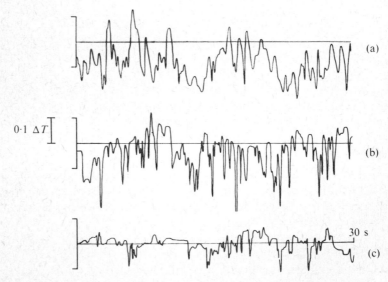

FIG. 8.6. Temporal variation at one place of temperature $\mathcal{T}$. $(Ra) = 10^9$. Depths below surface: (a) 0·1 mm; (b) 1 mm; (c) 10 mm. Equivalent width of record is 30 s.

1. The spatial variation obtained over a horizontal line shows wiggly records with a wide range of length-scales of variation, the sharpest being much narrower than the depth of the container. In the interior we note the clear correlation between θ and w, and in the sublayer between θ and u.

2. The temporal variation obtained at a fixed point over an interval of time has a more pronouncedly spiky nature in the interior.

These features suggest that plume-like elements are the dominant heat transport entity. We note especially the amplitude of the temperature fluctuations, typically of the order of $0\cdot1\,\Delta T$.

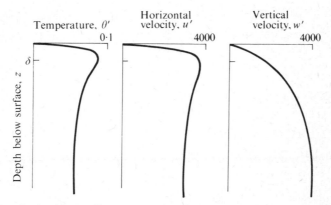

Fig. 8.7. Amplitude of fluctuations: r.m.s. values of temperature θ', horizontal velocity u', and vertical velocity w'. $(Ra) = 10^9$.

We are all familiar with this phenomenon, especially the case of the shimmering of sunlight over a flat surface on a summer's day. In Maori legend this is referred to as the 'dancing of Tane-rore', since she was the daughter of the sun and his summer wife.

The region at the outer part of the sublayer where the temperature fluctuations are a maximum may well be a region of extreme scattering of seismic waves and could produce the so-called seismic low-velocity zone.

The temperature spectra

1. *The temporal spectrum.* From measurements over an interval of time at a fixed point we can obtain a temporal spectrum. At high frequencies there is an extremely rapid fall in energy. This cut-off frequency is about where $\tau f \sim 1$, where $\tau = \delta^2/\kappa$, the time-scale of the sublayer. At low frequencies there is a much more gradual fall in energy and the precise details depend on the shape of the container. It is worth noting that even in the laboratory apparatus these low frequencies are very low indeed, being typically less than 10^{-3} Hz. In between these two extremes, even though individual

spectra vary somewhat, the spectrum is 'white': in other words, in this range all the spectral components are of similar amplitude. The white region in the laboratory typically covers a frequency range of about 30:1, with the bulk of the energy in a 300:1 frequency range.

2. *The spatial spectrum.* From measurements along a horizontal line over a very short interval of time we can obtain a spatial spectrum, the energy

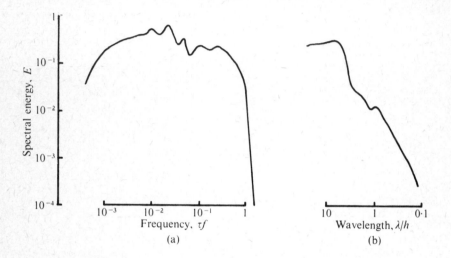

FIG. 8.8. Spectral energy E as a function of wavelength λ and frequency f: (a) temporal, (b) spatial (after Deardorff). Note that $\tau = \delta^2/\kappa$, where δ is the sublayer thickness. $(Ra) = 10^9$.

distribution over components of various horizontal wavelengths. This type of spectrum broadly shows the largest energies in the eddies of largest size, those that fill the container, with a fall-off roughly inversely proportional to eddy size. The spectra are somewhat depressed below the general trend where the horizontal wavelength λ is comparable to the depth of the container.

The velocity fluctuations exhibit very similar temporal and spatial spectra.

More than any other, the temporal spectra emphasize the most striking geological implication. Even if the earth were an otherwise homogeneous body made hot and allowed to cool at the surface, vigorous activity over time-intervals from $\tau f = 1$ to, say, $\tau f = 10^{-3}$ could be expected. These frequencies are of order 1 cycle per 10^8 yr to 1 cycle per 10^{11} yr. Energy is available fairly equally to all these frequencies but is dominantly in the eddies of global scale.

The heat transport process

Each of the individual processes which together produce the turbulent field can be measured. The relative level of each process gives important clues to the local mechanics.

1. As we have already seen, in part, the active sources of the field arise in the sublayer. This is the region of production and control of the vigorously fluctuating processes. We shall study this important region further, later in this chapter.

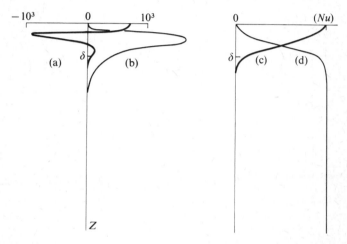

FIG. 8.9. Contributions to mean production and transport, as functions of depth z, below the surface at $(Ra) = 10^9$: (a) rate of production of θ^2; (b) molecular transfer rate; (c) vertical heat transfer rate by thermal conduction, $K\langle \partial \mathscr{T}/\partial z \rangle$; (d) vertical heat transfer rate by the turbulent fluctuations, $\rho c\langle w\theta \rangle$. Note the localization to the sublayer except for (d). ((a), (b) after Deardorff.)

2. But now we see another aspect of the turbulent field. Its importance cannot be overstated. That quantity $\rho c\langle w\theta \rangle$ represents a contribution to the mean vertical flux of heat. Below the sublayer the heat flux is transported solely by the processes represented by this term. It arises from the strong correlation of the vertical velocity fluctuations and the temperature fluctuations. This process is diminished only in the sublayer where, as the effectively rigid and isothermal surface is approached, both w and θ fall toward zero and the heat flux is increasingly carried by the process of thermal conduction.

The mean heat flux is transported through the interior solely by the interaction of the fluctuations of the turbulent field. The molecular processes of viscosity and thermal conductivity play no part whatsoever in this transport in the interior. Provided that the Rayleigh number is sufficiently large for the system to be turbulent the viscosity and thermal conductivity of the interior are irrelevant quantities. No deductions about the viscosity or thermal conductivity of the interior is possible from vigorous convective models of the earth's thermal history. These quantities are relevant only within the sublayer.

This situation is not peculiar to thermal turbulence. A precisely analogous situation applies for ordinary turbulence. There the quantity analogous to the heat flux is the stress which, as shown by Reynolds in 1894, is maintained in the interior solely by the interaction of the velocity fluctuations.

A meteorological perspective

A good example of a system normally described by these methods and familiar to all of us is the meteorology of the atmospheric system. If you measure the temperature (or humidity, pressure, and velocity) of the air at a fixed point you will find that it fluctuates irregularly with time. Similarly, if you measure the temperature (or humidity, pressure, and velocity) at a given time you will find that it fluctuates irregularly with position. Yet if you read a weather report or forecast much of this irregularity is not referred to. The various quantities mentioned are clearly suitable averages over a volume or an interval of time. Further, the major fluctuations are represented as discrete entities such as the large flat eddying motions of cyclones or anti-cyclones. For the purposes of compact description we pretend that all the different motions can be superimposed on an otherwise uniform system. In reality these components do not have such a separate existence; rather they are merely local intensifications of the fields generated by a continuous system in which all the parts have a high degree of mutual interaction.

Another aspect of systems like this tends to be obscured when descriptions are given solely in terms of mean field quantities. There is usually no indication of the mechanism by which the most important quantities of interest, namely the fluxes, are generated. In such systems these fluxes are dominated by the fluctuating quantities. For example, in the case of the thermal turbulence of a layer of fluid cooled from above, below the sublayer we find that the vertical heat flux is controlled solely by the interaction of the vertical velocity fluctuations and the temperature fluctuations, so that $f = \rho c \langle w\theta \rangle$ throughout the interior.

Thus, when later in this book I describe certain more-or-less discrete motions these are taken to be additional to whatever pre-existing motions there are. Just think of this in the same way as when you look at a weather map.

Development of the interior flow

The cellular pattern characteristic of laminar convection at low Rayleigh numbers remains up to Rayleigh numbers of the order of 10^6, but there are pronounced changes. Already, at a Rayleigh number of about 5000, the flow field in each cell is dominated by boundary layers on the walls and thin plumes in the middle of the cells. Further, as the Rayleigh number increases the cell boundaries are seen to be no longer stationary but to move slowly; occasionally cells are annihilated while others are formed. This tendency becomes very pronounced at Rayleigh numbers of the order of 10^6. We observe that near $(Ra) = 10^4$ the time of adjustment of cell boundaries is very long compared with the orbit time of a fluid particle. Near $(Ra) = 10^6$ these times are comparable. Nevertheless, the heat-transfer coefficient remains proportional to $(Ra)^{\frac{1}{4}}$, indicating that the motion is still laminar and is dominated by the combined processes of diffusion and advection in the boundary layers and plumes. Because of the now rapid drifting of the cells, measurements with temperature or velocity probes will indicate a strongly varying signal.

At a Rayleigh number somewhat above 10^6, however, we observe a remarkable and pronounced change in the system. Regular or quasi-regular cellular motion in the body of the fluid disappears, and fluid elements are seen falling from the cold upper wall in a random manner. The interior flow is dominated by a rapidly changing collection of eddying motions of a wide range of scales. It is as if the tight interaction of laminar flow between the boundary layers and the interior has been broken so that the boundary region and the interior are somewhat distinct.

The thermal sublayer

All evidence of 'cellular' motion is not, however, lost, for it is found to persist in the sublayers. If we view the layer from above, so that we look directly into the thermal sublayer, we see the horizontal structure and notice a definite pattern of motion *confined* to the sublayer. We see a polygonal structure, with flow rising in the interior of each polygon and plunging downwards in thin sheets at the margins of the polygons. There are three very important aspects of these motions:

(1) the pattern is continually changing, cell boundaries being annihilated, others forming, and upwelling sites changing position;
(2) the coherence of the pattern persists only within the sublayer, so that the motion is quite different from that in ordinary cellular convection;
(3) the horizontal length-scale of the polygonal structure is set solely by the length-scale of the thermal sublayer.

If we now observe the thermal sublayer as the whole system slowly cools down we find that the horizontal length-scale of the polygonal sublayer structures increases with time. We find that the Rayleigh number, based on the length-scale and the temperature difference across the sublayer, remains constant throughout the cooling. In other words, $(Ra)(\delta/h)^3 = (Ra)_c = 730$, that is, a constant. The result simply states that the sublayer is, in effect, in a state of marginal dynamical stability.

It is convenient to express the mean heat flux f in terms of the properties of the sublayer, where ϑ is the temperature drop across the sublayer, so that

$$f = \rho c \kappa \vartheta / \delta,$$
$$\delta = S\vartheta^{-\frac{1}{3}},$$
$$S \approx 9(\kappa \nu / \gamma g)^{\frac{1}{3}}.$$

Theoretical sketch: the mean temperature profile

It is useful to have a simple empirical expression for the mean temperature profile $T(z)$. We note, writing $y = z/\delta$, that:

(1) at $y = 0$: $\partial T / \partial y = \vartheta$, which is in effect the definition of δ;
(2) at $y = 1$: $T/\vartheta \approx 0.80 \pm 0.02$;
(3) at $y = 2$: $T/\vartheta \approx 0.97 \pm 0.02$.

A function which fits this data quite well is

$$T/\vartheta = 1 - \exp\{-(y + 0.6y^2)\}.$$

I shall use ϑ throughout this book as the characteristic temperature of the earth's interior. Note especially that the temperature gradient at the top of the sublayer is ϑ/δ, but the actual temperature reaches ϑ at a depth of about 3δ. Below this depth, the mean temperature is constant in the laboratory, but in the earth it will lie close to the adiabatic temperature. In that case it is necessary to be able to refer to the earth's mean temperature. The mean temperature of an adiabatic sphere is

$$\xi \equiv \tilde{T}/T_s \sim e^{-\beta h}(1 + \tfrac{1}{4}\beta a + \tfrac{1}{20}\beta^2 a^2 + \cdots), \quad \beta = \alpha g/c,$$

where T_s is the temperature at depth h. For our planet, $\xi \approx 1.25$ if ϑ is used as the reference temperature. Thus for $\delta \ll a$ the earth's mean temperature is 1.25ϑ.

Theoretical sketch: the mechanics of the sublayer

1. Consider first the following sequence of events. Suppose that a portion of the sublayer has been removed so that at a particular instant in a local region the temperature is uniformly ΔT from the interior right up to the surface. Thermal diffusion will thereafter begin to cool the fluid below the surface in an ever-thickening layer of extent $\delta \sim (\kappa t)^{\frac{1}{2}}$. This layer is at first stable even though more dense fluid is on top of less dense fluid, because the local Rayleigh number $(Ra)_* = \gamma g \vartheta \delta^3 / \kappa \nu$ is initially small. But as δ increases with time, so does $(Ra)_*$ until $(Ra)_* \sim 10^3$. The layer is then unstable and a rapid process of ejaculation of the colder denser fluid into the interior occurs, leaving the region locally denuded again of

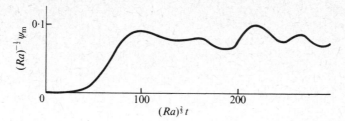

FIG. 8.10. Development of the flow from rest: maximum value of the stream function ψ_m, a measure of the total flow circulation per unit time, as a function of time. In dimensionless form. After an interval of 100 time units, during which the flow develops increasingly rapidly, the system is in statistical equilibrium.

FIG. 8.11. Time development of the proto-sublayer in a viscous fluid with $(Ra) = 10^9$. Stream function ψ and temperature $\mathcal{T}$ at dimensionless times $\kappa t/h^2$: (a) 0.5×10^{-4}, $\psi_m = 32$; (b) 10^{-4}, $\psi_m = 245$.

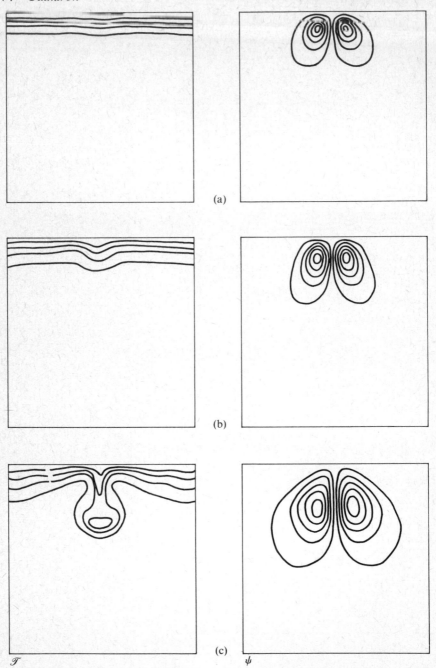

FIG. 8.12. Growth of an individual blob at $(Ra) = 10^9$ showing the entrainment into the blob at dimensionless times (a) 2×10^{-5}, $\psi_m = 15$; (b) 4×10^{-5}, $\psi_m = 120$; (c) 6×10^{-5}, $\psi_m = 700$.

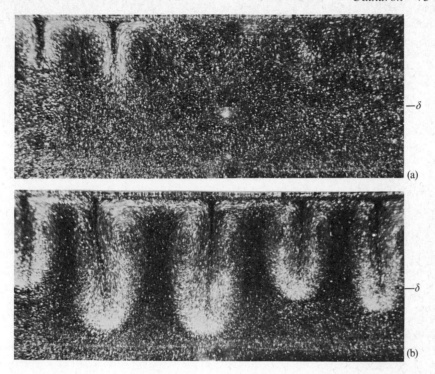

Fig. 8.13. Photographs of the development of the proto-sublayer following sudden cooling of a hot fluid layer from above. (a) growth confined to sublayer; (b) rapid entrainment.

its sublayer. This process is envisaged as occurring at random over the upper surface. Simple though the model seems, it is possible to calculate both the mean temperature and r.m.s. temperature fluctuations as a function of depth and we find quite a good fit with experimental data.

2. It is possible, however, to study in detail the processes in the sublayer by means of experiments on a proto-sublayer, where we start up a system from rest by suddenly cooling from above and look at the initial development of the layer.

For a hot layer of fluid at rest with its upper surface suddenly cooled at time $t = 0$, the development proceeds as follows.

(a) The upper portion is progressively cooled by thermal conduction to a depth $\delta \sim (\kappa t)^{\frac{1}{2}}$. There is negligible motion.

(b) Once the cooled layer reaches a thickness such that $(Ra)_* \sim 10^3$ the now unstable cooled layer overturns, and embedded in it we see a series of eddies of the same dimensions as δ. There is negligible motion in the uncooled interior.

(c) The amplitudes of the eddies grow roughly exponentially in time, reaching finite amplitude but remaining embedded in the still growing cooled layer. Large distortions of the hitherto vertically stratified temperature field are produced.

(d) Each eddy begins to rapidly entrain fluid from the surrounding cooled layer, bringing hot interior fluid closer to the upper surface.

(e) Now as virtually discrete and distinct entities, these blobs of cold fluid fall out of the thermal layer into the interior and stir it up.

Thus, throughout the interval of gestation of each element, it remains embedded in the (proto-)sublayer and only after it is sufficiently buoyant and vigorous to be free of the constraint of molecular processes in the (proto-)sublayer is it able to eject itself into the interior.

After the initial disruption of the proto-sublayer the process continues, effectively at random over the sublayer, and the over-all flow is already close to statistical equilibrium with the maximum circulation varying ± 20 per cent about its mean value.

The importance of the sublayer

This zone at the top of the mantle is the major active zone of the entire earth. It acts as a buffer between the well-mixed interior and the crustal systems.

It is important to emphasize that within the sublayer molecular processes—both thermal conductivity and viscosity—are important, whereas in the interior molecular processes play a negligible role in the global mechanics. In particular, therefore, the values of κ and ν appropriate to evaluation of this model are those of the material in the sublayer and *not* those of the interior. The single most characteristic parameter of this model is the length-scale $(\kappa\nu/\gamma g\vartheta)^{\frac{1}{4}}$ evaluated for the uppermost layer of the system: it determines the fundamental scales of length, time, and velocity for all macrogeological systems.

This sublayer scale is by far the single most important parameter in describing the interior dynamics of bodies of planetary size. It is the key to a description of all macrogeological systems.

The importance of the temperature fluctuations

I cannot emphasize enough the simple but far-reaching observation of the combined temporal and spatial variation of temperature in a vigorously convecting body. This variation is, as it were, an automatic built-in feature of the dynamics of such a system. The nature of this variation is most striking when we look at the spatial temperature variation at some given depth. For example, at the base of the thermal sublayer we have temperatures in the range, say $(2000 \pm 200$ r.m.s.$)\,°C$ with prominent length-scales of the order of

100 km and larger. These temperature variations led both to variations of density and to very large variations of viscosity, and although the mean properties change relatively slowly through geological time the time-scale of major alterations of the variations is of the order of 100 million years. These variations provide the essential link between processes on a global scale and the smaller-scale systems in the upper mantle and crust.

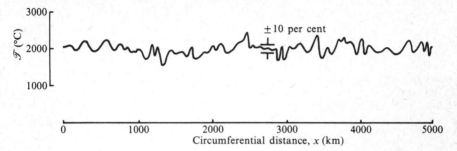

FIG. 8.14. Spatial variation of temperature at the base of the thermal sublayer in the earth. Schematic profile obtained from laboratory models. Temperature $\mathcal{T}$, in °C, at a given instant, at a fixed distance (about 100 km) below a point on the surface as a function of circumferential distance x, in km. Note that although the mean level and indicated r.m.s. range will change little in 100 million years the individual wiggles will be completely different.

9. Geological vigour

A thermal history

We now have more than enough information to discuss the rate of release of energy on a global scale throughout geological time. Here we are concerned that there is enough energy released at each stage throughout the life of the body to run the geological systems and to get some idea about the over-all rate of evolution of the reactor and the change in its vigour.

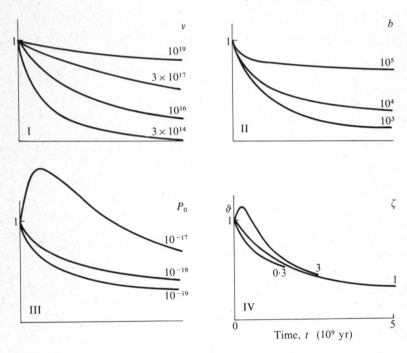

FIG. 9.1. Thermal histories. Characteristic temperature $\vartheta(t)/\vartheta(0)$ for the four types of model discussed in the text: I, constant viscosity; II, add variable viscosity; III, add radioactive heat; IV, add gravitational energy release. The key new parameter for each model is indicated.

We investigate a variety of zero-dimensional models where we in effect integrate the energy equation over the entire volume of the earth. The shape of the earth is required only to specify the ratio of surface area to volume.

Consider a hot fluid sphere losing heat because the surface temperature is maintained at a fixed temperature below that of the interior. The heat flow through the surface will be made up by contributions from the internal energy sources. Since energy must be conserved, the rate of loss of energy by the surface heat flux f through the surface area A will be balanced by the rate of loss of total internal energy, so that if V is the volume of the body of mean density ρ and E is the total internal energy per unit mass we have $\mathrm{d}(\rho V E)/\mathrm{d}t = -Af$. Here we consider only three sources: thermal energy, gravitational energy released during freezing, and radioactive decay. The whole story can now be put together.

TABLE 9.1

Cooling sphere model

At $t = 0$; given: $\vartheta = \vartheta_0$

(1)	$P = P_0\,\mathrm{e}^{-\lambda t}$	Radiogenic power
(2)	$T = T_0 + \vartheta$	Absolute temperature
(3)	$v = v_0\,\mathrm{e}^{-b/T}$	Kinematic viscosity
(4)	$S = 9(\kappa v/\gamma g)^{\frac{1}{3}}$	Sublayer scale
(5)	$\delta = S\vartheta^{-\frac{1}{3}}$	Sublayer thickness
(6)	$f = \rho c\kappa\,\vartheta/\delta$	Surface heat flux
(7)	$\dfrac{\mathrm{d}r}{\mathrm{d}t} = -fr/\rho\mathscr{H}a$	Core radius
(8)	$\dfrac{\mathrm{d}W}{\mathrm{d}t} = W_0\left(\dfrac{r^4}{a^5}\right)\left(\dfrac{\mathrm{d}r}{\mathrm{d}t}\right)$	Gravitational energy release
(9)	$\xi\dfrac{\mathrm{d}\vartheta}{\mathrm{d}t} = -\dfrac{3f}{\rho ca} + \dfrac{1}{c}P + \dfrac{\mathrm{d}W}{\mathrm{d}t}$	Energy balance

NOTES

(a) Decay constant $\lambda = 5\cdot4 \times 10^{-10}\,\mathrm{s}^{-1}$; initial radiogenic power $P_0 \sim 10^{-19}\,\mathrm{W\,kg}^{-1}$; surface temperature $T_0 \approx 300\,\mathrm{K}$; viscosity coefficient $b \sim 10^4\,\mathrm{K}$; reference viscosity $v_0 \sim 10^{15}\,\mathrm{m}^2\,\mathrm{s}^{-1}$; enthalpy change on freezing $\mathscr{H} = 1700\,\mathrm{kJ\,kg}^{-1}$; gravitational energy release taken in the form $W_0 = \zeta c\vartheta_0$ with $\zeta \sim 1$, $K = 4\,\mathrm{W\,m}^{-1}\,\mathrm{K}^{-1}$, $\rho = 4\,\mathrm{g\,cm}^{-3}$, $c = 1\,\mathrm{kJ\,kg}^{-1}$, $\kappa = 10^{-6}\,\mathrm{m}^2\,\mathrm{s}^{-1}$, $\vartheta_0 = 3000\,\mathrm{K}$, $\xi = 1\cdot25$.

(b) A single radioactive source is used (see Chapter 5).

(c) The gravitational source term uses the model of Chapter 5 where $W \propto r^4$. Strictly W should be evaluated in detail (see Chapter 6), but this refinement alters the results only slightly.

(d) The minor assumptions are: g is approximately constant across the sublayer: g is approximately constant in time; $\delta/a \ll 1$; all properties of the material are fixed except for v.

Convective model I: *constant properties, internal energy only*

In the simplest possible case: $P_0 = 0$, $W_0 = 0$, $b = 0$; there is no source of radiogenic or gravitational energy, and the material is homogeneous and of constant properties. The energy equation can be integrated immediately.

Using the relations for f, δ for the sublayer we deduce that

$$\frac{\kappa t}{a^2} = \left[\frac{(Ra)_{\mathrm{c}}}{(Ra)_0}\right]^{\frac{1}{3}}\left\{\left(\frac{\vartheta}{\vartheta_0}\right)^{-\frac{1}{3}} - 1\right\}$$

for the time t to reach an interior temperature ϑ, given at $t = 0$ an initial temperature ϑ_0 and a Rayleigh number $(Ra)_0$ and where $(Ra)_c \approx 730$. Notice that this expression is in dimensionless form.

Let us keep all the parameters fixed at their nominal values and see how things change with different choices of ν in the range 10^{14}–$10^{20}\,\mathrm{m^2\,s^{-1}}$. Clearly, with $\nu \sim 10^{14}\,\mathrm{m^2\,s^{-1}}$ the body cools down very quickly, whereas with $\nu \sim 10^{20}\,\mathrm{m^2\,s^{-1}}$ it has so far cooled very little. We notice that the model has

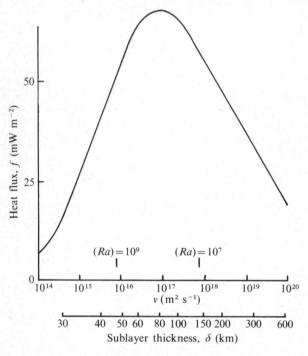

FIG. 9.2. Heat flux in $\mathrm{mW\,m^{-2}}$ at $t = 5 \times 10^9\,\mathrm{yr}$ for given ν, model I, constant properties. The derived values of the sublayer thickness δ are shown on a separate scale. The Rayleigh numbers 10^7, 10^9 are indicated.

the interesting property of a maximum in $f(\nu)$, the heat flux (at a given time) as a function of ν. All the models to be discussed have this feature. For a cooling time of 5000 million years this maximum occurs near $\nu \sim 10^{17}\,\mathrm{m^2\,s^{-1}}$. We can choose ν to give a good fit to present-day data. The constant-property model is in fact sufficiently good for most purposes. It is by far the outstanding result presented in this book and allows us to make valuable statements about a wide range of problems in planetary development. But it is not perfect. For example, a value of $\nu = 3 \times 10^{16}\,\mathrm{m^2\,s^{-1}}$ gives a satisfactory heat flux and core radius, but rather a thin upper mantle. To improve the model, we need to thicken the upper mantle, and at the same time slow it down.

Convective model II: *add variable viscosity*

We want to decelerate our first model. This would be readily achieved by having it stiffen up as it cooled down. But this is precisely what a viscous fluid will do. I have already noted in dealing with the rheological behaviour of solids that $v \approx v_0 \exp(b/T)$.

The effect of allowing v to vary with temperature is readily understood. Suppose b were very large. Then as soon as the temperature falls a small amount the current viscosity becomes very large; hence the Rayleigh number becomes very small and the convection is turned off. Thus, for $b \to \infty$ convective cooling is negligible, and since we have also ignored ordinary thermal conduction, the body does not cool at all. At the other extreme, when $b = 0$, convection proceeds at its greatest rate and the body cools quickly. Between these two extremes, in the real world as it were, a moderate cooling rate is achieved.

With b in the range 10^3–10^4, we have good estimates for the upper mantle thickness and the core radius, but the heat flux is rather low. What we have done is simply to use up our energy too early when the viscosity was much smaller. Clearly we now need some more energy to 'top up' our model.

Convective model III: *add radiogenic heating*

Radiogenic heating is the energy booster most commonly invoked in present-day theories. There is no problem in getting enough heat flux. For example, with $P_0 = 10^{-18}\,\mathrm{W\,kg^{-1}}$, we already have more than enough. Larger values lead to a phase of heating up, since the rate of generation exceeds the ability of the convection to transport the energy. But all these values lead to poor estimates of core radius and upper mantle thickness. A value somewhat greater than $10^{-19}\,\mathrm{W\,kg^{-1}}$ gives about as good a fit as possible.

Convective model IV: *add gravitational release*

When we finally take into account the gravitational energy available from the rearrangement of the interior as the core freezes out it would seem as if we now have an embarrassing overabundance of energy. Fortunately the rate of gravitational energy release is rapid at the beginning and is almost negligible after 5000 million years. For $\zeta = W_0/c\vartheta_0 \gtrsim 1$ there is an early phase of heating up. Over the range of ζ near values estimated in Chapter 5 the details do not differ much. The best fit indicated is for $\zeta \sim 1$

Vigour

As our planet evolves through geological time, the rate of the internal processes changes. How can we most simply characterize the state of activity?

TABLE 9.2
Cooling models

Model I: Constant viscosity ($b = 0$, $P_0 = 0$, $\zeta = 0$)

v (m^{2}s^{-1})	3×10^{18}	10^{20}	3×10^{20}	3×10^{21}	10^{23}
f	13	54	64	63	33
r	3000	3400	3700	4500	5500
ζ	29	49	62	110	310

Model II: Variable viscosity ($P_0 = 0$, $\zeta = 0$)

b (K)	10^3	10^4	10^5
f	26	25	12
r	3200	3700	5100
δ	48	150	780
v/v_0	4·5	370	$1·4 \times 10^5$

Model III: Add radiogenic heating ($\zeta = 0$)

P_0 (W kg^{-1})	10^{-19}	10^{-18}	10^{-17}
f	30	72	420
r	3500	2500	40
δ	130	70	22
v/v_0	270	56	3·1
η	0·13	0·85	3·0

Model IV: Add gravitational energy release

ζ	0·3	1	3
f	32	35	38
r	3400	3000	2000
δ	120	120	110
v/v_0	240	210	180
η	0·13	0·12	0·11

Data from four models, specified in Table 9.1, now, after 5×10^9 years, each progressively more elaborate. The following quantities have the standard values and units, unless otherwise stated: $v_0 = 10^{15}$ m^2 s^{-1}; $b = 10^4$ K; $P_0 = 10^{-19}$ W kg^{-1}. The tabulated values have the units: heat flux f, mW m^{-2}; core radius r, km; upper mantle scale δ, km; viscosity is given as a proportion of v_0. $\zeta = W_0/c\vartheta_0$ is a measure of the energy available from gravitational release as a proportion of internal thermal energy. In model III and IV, $\eta =$ the power output from radioactive decay as a fraction of that from the internal thermal energy.

From the point of view presented here, there is one simple and dominant parameter which tells us unequivocally about the internal vigour: the roll-over time-scale of the upper mantle. This is the time required for a portion of the upper mantle to be renewed. As we shall see in later chapters, all the macrogeological systems are controlled by the upper mantle. We now estimate the roll-over time $\tau \approx 0.1\,\delta^2/\kappa$, where δ is the upper mantle thickness and κ the thermal diffusivity. For model I, writing $\xi = \{(Ra)_0/(Ra)_c\}^{\frac{1}{3}}$ we have $\delta/a = (1 + \xi\kappa t/a^2)/\xi$, so that δ grows linearly with time from its original value $1/\xi$, which is, with $v_0 = 10^{15}\,\mathrm{cm^2\,s^{-1}}$, about 25 km. Thus the roll-over time-scale increases quadratically with time. The geological vigour of our planet has already diminished by at least an order of magnitude.

The rate of all global processes will be proportional to the roll-over time: this includes polar wandering and all macrogeological crustal rearrangement. Strictly we should also allow for the changes in the fluctuating velocity- and temperature-scales as appropriate to each process, but these scales change very little with Rayleigh number (see Table 8.1, p. 64).

A planetary perspective

The concept of geological vigour is best appreciated by comparing our earth with other bodies of planetary size. We use results from the convective models just discussed. For example, if we compare the predicted roll-over times for the earth, the moon, and Mars, we find that the moon is almost dead, Mars is getting tired, but our globe is still rather youthful.

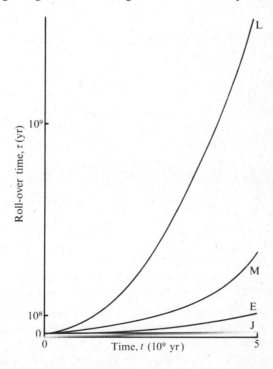

FIG. 9.3. Roll-over time τ of the upper mantle as a function of time t of some bodies of planetary size: L, the moon; M, Mars; E, the earth; J, Jupiter.

When we look at all the planets we see that the earth lies between the giant planets, which have hardly cooled at all and are still extremely vigorous, and the so-called terrestrial planets, the smaller of which are markedly slowed down. To some degree we see here the over-all geological story of our planet at a glance: after 10^7 years, like Jupiter (apart from the methane, of course?); after 10^{10} years, like Mars; after 10^{20} years, like the moon.

TABLE 9.3

Planetary evolution now (bodies are listed in order of size)

	δ	ϑ	f	τ	v
Moon	260	790	8	210	0·7
Mercury	190	830	20	110	0·8
Pluto	170	900	30	87	0·8
Mars	170	1150	20	93	1·1
Venus	120	1440	54	45	1·6
Earth	116	1440	58	42	1·6
Neptune	89	2330	38	25	45
Uranus	98	2420	30	30	50
Saturn	96	2750	16	29	75
Jupiter	71	2720	46	16	72

NOTES

Sublayer thickness δ (km); characteristic temperature ϑ (K); heat flux f (mW m^{-2}); roll-over time τ is in units of 10^6 yr; core volume v as a percentage of total. Note: δ_0 is the same for all planets, about 25 km for standard parameters. The only parameters special to each body are: radius a and surface gravity g. Model I with: $(Ra)_c = 730$, $\gamma = 10^{-5}$ K^{-1}, $\vartheta_0 = 3000$ K, $\kappa = 10^{-6}$ m^2 s^{-1}, $v = 3 \times 10^{17}$ m^2 s^{-1}, $c = 0.84$ kJ kg^{-1}, $\mathscr{H} = 700$ kJ kg^{-1}. Note this gives for the earth $(Ra) = (Ra)_0 = 2.6 \times 10^8$. I have chosen to use the simple Model I for this tabulation, rather than the better model IV, since the table entries are readily verified using a slide rule or values recalculated for other parameters. As a result, the values of both $\mathscr{H}$ and here (Ra) need to be rather smaller than they should be. The value of $\mathscr{H}$ has been chosen so that the core radius of the earth is its actual value. Although there is a rather wide range of choice of parameters they all lead to the same over-all result as presented here.

Commentary

Until very recently, all thermal histories have been conduction models. These models are of negligible geological interest, because they allow release of energy from only the upper few hundred kilometres of the mantle. At the time they were the only models that could be handled analytically or numerically, but that is no longer the case. Further, implicit in the present model is the assumption that the earth was formed hot and has subsequently been cooling down. It seems to me that models of a cold origin ignore the geological observations of the persistent mobility of the earth.

In the present turbulent model, with access to all the internal energy but with the impedence of the very viscous upper mantle, we have sufficient energy to run the earth. From our chosen point of view we are forced to regard the sources of radiogenic heat as producing local thermal anomalies rather than providing the bulk of the earth's heat loss. This viewpoint is quite different from that of conduction models. In those the role of radiogenic heat is vital since, as we have seen, conduction alone provides access to a very small amount of the earth's internal energy.

Studies of the earth's thermal history have been bedevilled by pre-occupation with radiogenic heat sources. How has this come about? Kelvin's famous conduction model of a cooling earth fits the known surface heat flux with an age of about 10^8 years, a result rightly ridiculed by the geologists (who seem ever since to have been suspicious of the physicist). Clearly the model needs more energy. As chance would have it, radioactivity (rather than 'thermal turbulence') was discovered shortly afterwards and gave a possible additional energy source—the great store of internal energy in the Kelvin model being neglected. The conduction–radiogenic model still proved adequate for a time, but it founders immediately on the difficulty that there are roughly the same regional heat fluxes on continents as in the oceans, in spite of the powerful lithophilic tendency of the principal radiogenic elements.

Contemporary conduction models have tried to reach deeper into the mantle by invoking the weak increase of thermal conductivity with temperature. The convection–internal-thermal-energy model reaches all the earth's thermal energy. Nevertheless, there is an interesting parallel with an early conduction-model problem. There it was soon found that, if typical granitic radiogenic concentrations were assumed, nearly all the radiogenic material must be in the outer 100 km of the mantle. This layer would form a warm blanket over the thermally inaccessible interior. Here, however, if the viscosity of the mantle were uniform, and the earth were originally in a vigorous convective state, it would have cooled down rather too quickly. Our 'blanket' is the upper mantle in which the viscosity is high enough to give a thick layer dominated by conduction.

10. Global displacements

UP TO NOW we have concentrated our attention on average global properties below any given point on the surface, all such points being regarded as statistically the same. Before we leave the global scale let us look at the surface, not in section but in plan.

Tracking the crust with fossil magnetism

The year 1600 saw the publication of William Gilbert's momentous work in which he showed that the earth was a magnet. Since that time two outstanding observations about the magnetism of our planet have led to our present understanding of the global kinematics of the crust.

1. Rock-substance taken above and then below the so-called Curie temperature, typically about 700 °C, preserves a copy of the magnetic field at that time. Thus, even if a piece of matter has been moved about, it is possible in part to state that a displacement has occurred and where it was at a previous time just from its direction of magnetization. In particular, the apparent positions of the magnetic pole obtained from measurements of igneous rocks from various continents show a wide variety of pole positions, but for each continent these follow a continuous path unique to that continent.

2. The magnetism of the earth, which can be likened to that of an electric dynamo,† is not steady, but alternates so that occasionally it reverses its direction of magnetization. It remains in one state for periods of the order of 100 000 years and switches over in a period of the order of 1000 years. Thus, a rock sequence, a suite of rocks produced over an interval of time and laid out in space in an orderly fashion, preserves in its spatial megnatic variation a replica of the magnetic alternations that have occurred in the corresponding interval of time. This magnetic time-sequence is now known for about 80 million years into the past and can best be identified in lava-flow sequences or in the steady sedimentation on the ocean floor. In addition, a most remarkable situation is found in oceanic basins. There the magnetic pattern is striped: a sequence of bands side by side, alternately magnetized. This suggests that the material of these parts of the ocean floor has been produced in a narrow zone and subsequently displaced sideways.

† A practical dynamo, while normally regarded as a device for converting work into electric current, can also be considered as a dynamical method of producing a magnetic field.

Both these effects show that relative displacements of parts of the earth's crust have been going on more or less continually throughout geological time at rates that are typically $1-10 \, \text{cm} \, \text{yr}^{-1}$.

Analysis of relative crustal displacement

Once we recognize that gross redistribution of the matter of the earth is, and laid out in space in an orderly fashion, preserves in its spatial magnetic observation platforms from which we measure the mutual displacements of the earth's parts), we see the difficulty of finding a suitable frame of reference. All that we seem to have as an apparently permanent reference is the axis of rotation. Certainly the axis may change its rate of precession and over-all orientation measured relative to the solar system, but the time-scale for such processes will presumably be long compared to the time-scale of global geological processes, which is of the order of 10^8 years. Given then the axis of rotation as a reference, a crustal rearrangement relative to the axis of rotation can be analysed into three distinct components.

1. *Polar wandering*: displacement of the poles of rotation relative to the entire earth's crust considered as a single body.

2. *Translation*: a uniform lateral movement of a portion of the crust. On a sphere such a movement can be represented as a rotation about an axis outside the moving portion of the crust.

3. *Rotation*: turning of a portion of the crust about an axis passing through the rotating element.

This method of description of the crustal rearrangement does *not* mean, for example, that the crust actually moves as a single shell, quite the contrary. Think of it like this. We are familiar with both direct and alternating current. Imagine a circuit in which a current is alternating about a non-zero level— many of the circuits in a radio are of this type. It is still both convenient and practical to refer to the component of direct current and the component of alternating current in such a circuit, even though in practice their joint effect may be somewhat different from the sum of their effects when acting separately. Again, imagine a set of photographs obtained in the following way. A snapshot camera with a wide-angle lens is suspended from the ceiling of a large room and arranged to take photographs of objects below it. The room is crowded with people dancing old style. A flash photograph will show no movement at all. A photograph taken with the shutter open 1 s will show that the objects are actually moving but it would be difficult to say much about the form of the movement. A photograph taken with the shutter open 30 s will show many blurred but otherwise distinct traces revealing localized eddying motions. Finally, a photograph in which the shutter was

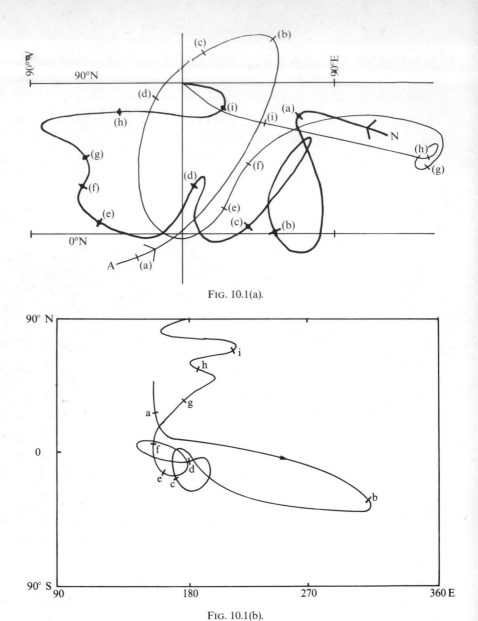

FIG. 10.1(a).

FIG. 10.1(b).

FIG. 10.1. Relative displacements of the continental crust. (a) Polar wandering curves, in present day coordinates, since 2750 million years B.P., for the continents of North America, N and Australia, A. The instants (in 10^6 yr B.P.): 2500, 2000, 1500, 1000, 500, 400, 300, 200, 100 are labelled a–i. (b) Polar wandering component common to all continental elements. Obtained from paleomagnetic data on apparent pole position of Africa, Antarctica, Australia, China, Europe, India, North America, Siberia, and South America. (c) Amplitude of rate of displacement relative to the mantle, in deg per 10^6 yr $(= 11 \text{ cm yr}^{-1})$, for the common polar component as shown in (b) and for individual elements relative to the common component.

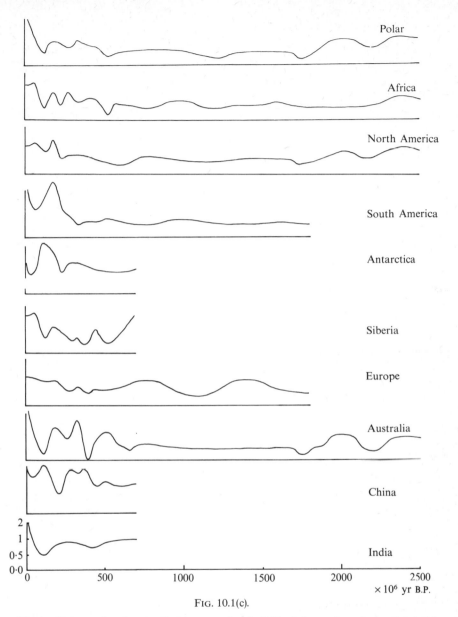

FIG. 10.1(c).

Obtained by assuming that the displacement of an individual element is made up of a global component together with a random component. Averages taken with weighting proportional to element area. Note the non-uniform vertical scale for which the ordinate is proportional to the square root of the rate. The graphs are in order of the areas of the elements. The apparently muted character of the curves before 500 million years B.P. arises from sparseness of data. These graphs can be used to obtain only a schematic picture of the sort of crustal rearrangement that has been going on. The original data is far too sparse and dubious to be able to rely on individual features. (Original paleomagnetic data obtained from a compilation by Seyfert and Sirkin 1973.)

open for 1000 s will show just a single eddy filling the area of the floor. We might interpret this as a result of a more or less rigid rotation of a single shell.

There is another important aspect of our observations of ballroom dancing. The amplitude of the global motion is of the same order as that of the intermediate scale eddies. So it is with the crustal dancing of the earth. The amplitude and rate of both the polar-wandering component and the components of smaller scale is of the same order. In these circumstances analysing a set of imperfect records of crustal movement into their components is especially difficult.

Polar wandering

Drift of the entire globe relative to the axis of rotation occurs continually and rapidly. Drift seems to have gone on throughout geological time and is at present going on at the angular rate of 90° in about 400 million years. This could arise from quite small redistributions of matter within the earth's interior.

Imagine the earth to be a fluid top, spinning in space. Suppose that it has been left undisturbed for a long time so that any internal readjustments are complete. Now the key quantity in the description of the mechanical state of such a body is its moment of inertia about axes through its centre of mass. Regardless of the axis we have a moment of inertia, and about different axes we will normally get different values. For example, if the body is corpulent the moment about an axis perpendicular to the corpulence will clearly be larger than about any other axis. Let us choose three mutually perpendicular axes, say, x, y, and z, about which the moments are A, B, and C such that $A < B < C$, where C is the largest one. The question is then, what is the relation of the axis of rotation to the disposition of the moment of inertia? Nothing could be simpler: the axis of rotation will lie along z, the reference axis about which the moment of inertia is greatest. Suppose it were otherwise, then the rotating body will experience torques—in the same manner in which a piece of unbalanced rotating machinery does—which, because our top is fluid, generate internal motions to redistribute the internal matter and by progressively deforming the body shape move the moment of inertia until the two axes coincide.

The ability of the shape deformation to follow the movement of the inertia axis is determined by the viscosity of the interior. For a homogeneous body the time-scale of this adjustment is determined, in a way similar to that for the bobbing up and down of the floating continents, to be $\tau = 19v/2ga$. With $v \sim 10^{17}\,\mathrm{m^2\,s^{-1}}$, $\tau \sim 10^3\,\mathrm{yr}$, a time that is very short compared to the time-scale of 10^8 years for the movement of the axis: the deformation is quite quickly performed.

Thus the body spins about the axis of greatest moment of inertia. Any

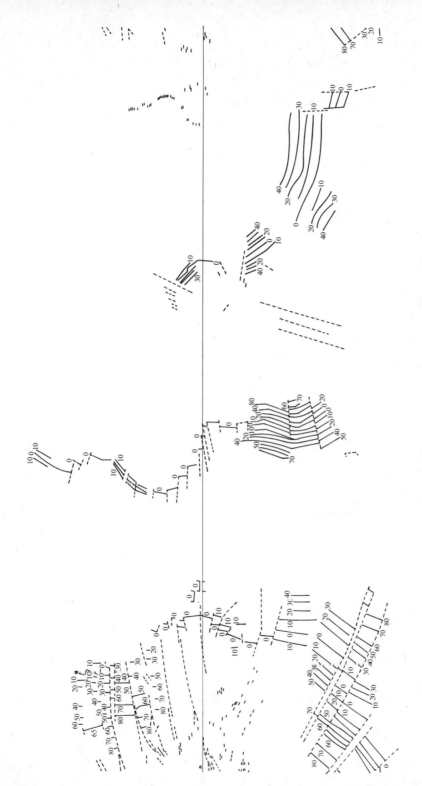

Fig. 10.2. Ages (in millions of years) of magnetic stripes on the ocean floors.

redistribution of matter induces a global circulation to maintain the axis of greatest moment of inertia as the spin axis. This is really the equivalent of lifting oneself by one's bootstraps on a global scale.

Departure of the interior from hydrostatic equilibrium

The detailed mapping of the geoid and the less detailed study of the global heat-flux pattern shows one very striking thing. There is negligible correlation

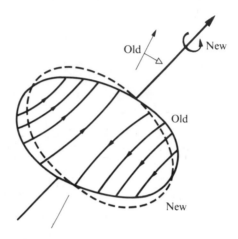

FIG. 10.3. Diagram for adjustment of orientation of moments of inertia by motion relative to the axis of rotation.

with the arrangement of the crust! Thus the departures from equilibrium arise deep in the interior. It is therefore both realistic and useful to consider the problem of global displacements arising from global processes, namely, polar wandering, as distinct from the meso-scale problem of displacements on a continental scale. Indeed, we find that quite different mechanisms operate on the two scales.

Theoretical sketch: random wandering

Because the contribution of crustal inhomogeneities is negligible in present non-hydrostatic moments of inertia we deduce that the masses bobbing about which lead to polar wandering are not crustal. Thus we envisage a statistically steady system of globally random density inhomogeneities, each of which has a typical lifetime. Presumably these elements are mainly those in the sublayer as described in Chapter 8.

Consider a maximum of n such elements, created at random times, of mass m_i, $i = 1, \ldots, n$, a function of time t such that, at creation, $m_i(t) = \xi m$, where ξ is a random number in the range ± 1, m is a fixed quantity, and $m_i(t)$ is

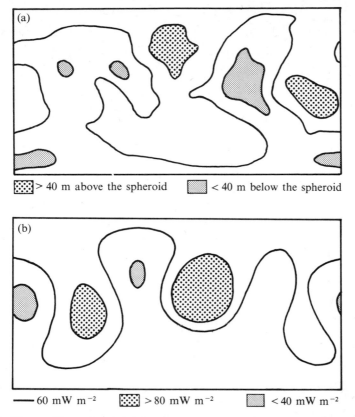

(a)

▓▓ > 40 m above the spheroid   ░░ < 40 m below the spheroid

(b)

——— 60 mW m⁻²   ▓▓ > 80 mW m⁻²   ░░ < 40 mW m⁻²

Fig. 10.4. Direct evidence for global interior processes. Equiangular projection of earth's surface, map centre 0° E, 0° N; highly distorted near the top and bottom lines representing the north and south poles. (a) The geoid relative to a flattening of 1/298·25. Positive values (heavy stipple) correspond to denser matter below. (b) Heat flux. These data are of extremely coarse resolution.

a diminishing function of time such that at $t = \tau$ it is destroyed. Here I choose $m_i(t) = m_i(t_0)\exp\{-3(t-t_0)/\tau\}$ for $(t-t_0)/\tau \leqslant 1$ and then delete m_i, so that at a later time a further element can be created.

The track of the principal axis of inertia is analogous to the problem of the so-called 'drunkard's walk'. There, for the simplest case we imagine an inebriated person leaving a bar located in the centre of a city with a rectangular street-pattern. Given that at each corner the drunk selects the next street at random, what will be the path? We find that on average the net distance travelled increases as the square-root of time. Our polar path differs in two main ways. First, it is on a sphere so that there is a tendency to come back to where we started from. Second, every now and then the polar 'drunk' is given a powerful nudge in a particular direction. This is

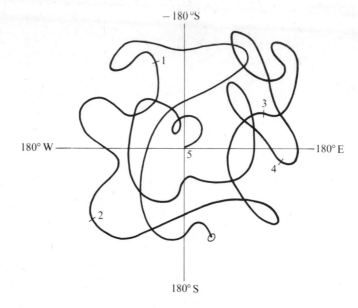

FIG. 10.5. Simulated polar-wandering curve from $0-5 \times 10^9$ yr. Position relative to the present north pole, the centre of the diagram.

because of the rapidity of the internal viscous flows in response to the out-of-balance torques. It is as if our polar 'drunk' were walking on a surface which was tilted occasionally to rather steep angles and the tilting directions changed rapidly. It would be a nightmare experience!

11. Crustal retreading

TILL NOW I have considered processes which affect the globe as a whole. Now we look into what is going on in the very small crustal vats. Because they are so small not only is the transit time of rock-substance through them generally small but they can quickly be reconstructed. But even that is not their most striking role. Look at it this way. So far we have constructed a basic model, a global model, which has the same properties below every point on the surface. If we now add a crustal system how will this affect the structure of the basic system? They key idea is to look for the *interaction* between the two systems. Adding the crustal system will alter the structure of the mantle, which will then have a different effect on the crustal system. We are dealing with a situation with a high level of mutual interaction.

The first step in this approach is in effect to add to the earth a uniform granitic crust which completely covers its surface.

Original segregation of oceanic and continental crust

How do we represent this crust? The essential feature to be kept in mind is the variation of viscosity both due to changes of composition or temperature. How can we model the expected large decrease in over-all viscosity with depth? A first step could be a two-layer system in which a thin, more viscous upper layer is placed on top of a deep less viscous layer. In the laboratory model the stiff layer is made of aluminium powder spread uniformly to a depth of about 0·1 mm on top of a convecting layer of medicinal paraffin. The question then arises, how will the presence of the initially uniform stiff layer affect the situation?

If the convection is rather weak, with small Rayleigh number, no substantial modification of behaviour is found. But if the convection is sufficiently vigorous the upper layer becomes violently disrupted. We identify the regions where the upper layer has been swept away as new oceanic floor and the remaining portions of the stiff layer as continents.

The role of a surface layer, simulating old crust, is partly to inhibit the motion in the sublayer beneath it by lowering the cooling rate. If the surface layer is sufficiently thick the outgoing horizontal viscous stress during the local disruption of the underlying sublayer may not persist for long enough to rupture the surface layer. Once the surface layer has been ruptured, however, there is a tendency for the crust to be stretched by a surface flow that is predominantly outward so that new blobs can more readily rupture the surroundings.

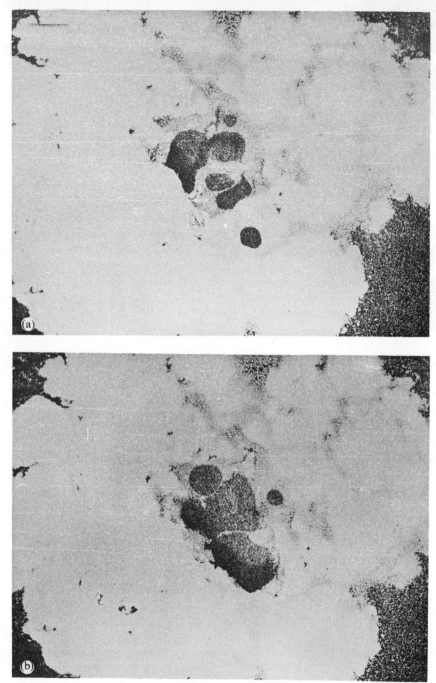

FIG. 11.1. Laboratory simulation of disruption of a supercontinent. Photographs of the upper surface of a layer of vigorously convecting fluid cooled from above on which is a thin very viscous surface layer. Time increasing from (a) to (d).

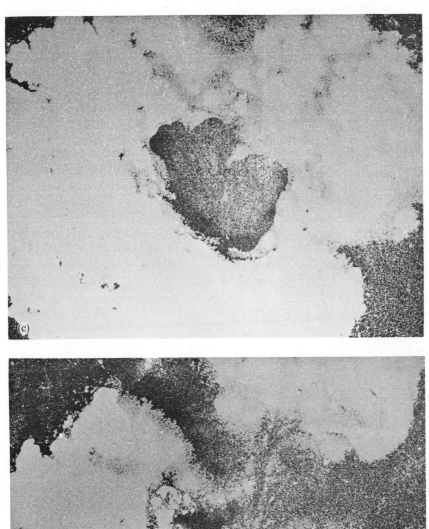

(c)

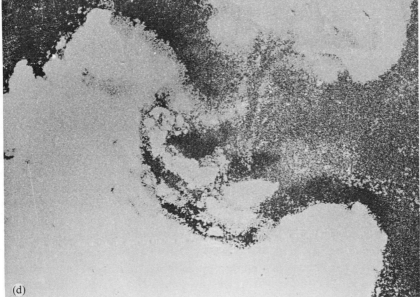

(d)

The model reveals an important aspect of the process of production of new oceanic crust. All but the smallest areas are produced, not by the action of a single buoyant blob, but by many. As each blob approaches the surface, the outgoing horizontal viscous stresses push outward the old crust. The net effect of the mutual action of many blobs is to expand a region progressively by sweeping away old crust at the margin. Each blob rising near a margin thereby produces a 'bay' in the old crust. Here I want to emphasize that the size of the buoyant elements is reflected, not in the size of the oceanic area (which is determined by the persistence of upwelling), but in the size of the embayments at the margin of the oceanic area. A further feature seen in the photographs, in which neither the camera nor the apparatus have been moved, should be noted: the new continental masses have been both translated and rotated.

Some continental remnants take the form of isolated or connected chains, which are usually marginal to the oceanic areas. These systems are analogous to the so-called island-arc structures. Where mantle convection has been locally persistent, fragments of continental-type material have been swept together and possibly augmented by discharge from the mantle itself. This process is seen in the model; for example, the strip of crustal material stretching from the right across the middle of the ocean floor of (b) in Fig. 11.1.

Thus even if the original crust were the same everywhere on the surface it would soon be broken up into a number of distinct parts of two kinds: *granitic crust*, either remnants of the original crust or pieces fabricated by joining together such pieces; *oceanic crust*, which has been swept clean of granitic crust. All this would take place in an interval of the same order as the upper mantle roll-over time, which is now about 10^8 years, but at the beginning of geological time is about 10^7 years.

A long period of complete crustal inundation would not be possible in these circumstances. Portions of granitic crust, would soon be elevated above sea-level, especially along the seams where granitic pieces are joined together. Erosion would then be able to operate and the deposition of typical sediments, graywackes and the like, would commence.

This model illustrates the really nice thing about such dynamical systems. We did not have to force the system in any way at all to break up the original crust: it is built into the natural dynamics of the system. The secret of modelling of this kind is to search for the minimum specification that is required for a system to develop in a particular way.

Flow near edge of a stationary continent

Having segregated our crust we recognize the possibility of the continental mass directly affecting the processes at depth. If we hypothesize that the continental material has a higher thermal resistance than the material in

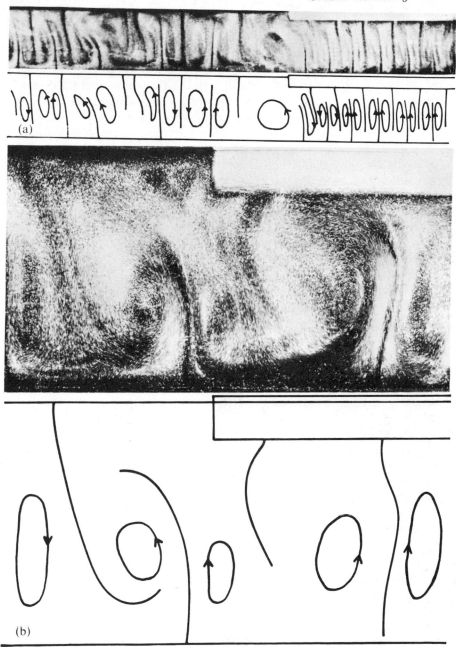

FIG. 11.2. Laboratory simulation of convection in the upper mantle near a stationary continental margin: (a) general view; (b) detail. The sketches show the eddy pattern and direction of fluid motion. Slab held fixed relative to the mantle.

which it is embedded, laboratory models show high heat flow in the oceanic region adjacent to the continent. The continental mass is analogous to an electric blanket, preventing to a degree the release of heat from the depths. This can be modelled simply by floating on part of the upper free surface of the working fluid a slab of wood to simulate the continental mass. Away from the continental edge, both in the oceanic areas and under the continent, there is a nearly regular, steady sequence of eddies. Those under the continent are thinner, but near the edge, and particularly for a considerable distance on the oceanic side, the motion is irregular and unsteady with a strongly pulsating upwelling of hot fluid under the edge. This suggests that this region would experience intense volcanism.

In these models, however, the continent is considered to be fixed. Now remove this restriction and allow the continental mass as a whole to move freely.

Crustal migration

Contemporary discussion has been concentrated on the possible movement relative to the mantle of masses of continental size. Further, these discussions have largely confined themselves to a 'drift' hypothesis. In brief, that model assumes the existence of streams in the upper mantle on which the mass rides. If the streams do not exist the mass does not move. The choice of the word 'drift' is therefore appropriate. There is, however, the alternative possibility of 'propulsion'. The essence of the propulsion hypothesis is that the crustal slabs are self-propelled. The difference between the two models is simply the difference between a raft and a raft with a motor.

A self-propelled sheet

Consider therefore the motion of a thin sheet floating in a layer of fluid heated from below where the layer itself may be convecting. If the sheet is homogeneous and its plan form is symmetrical (let us for the moment consider a rectangle) there is no preferred direction of motion. If, however, we increase the thermal resistance on one side of the rectangular sheet, say by local thickening, the impeded heat requires additional convective movements of the mantle for its removal. It is clear, as we shall see in detail below, that these movements provide a net horizontal thrust on the sheet. Thus, if the thermal resistance of the sheet to the vertical flux of heat is not centro-symmetric, motion of the sheet must ensue.

How can we most simply specify the geometry and thermal properties of such a sheet? In general we should state the thermal resistance of the sheet to vertical heat flow as a function of horizontal position on the sheet. For the purpose of this study we consider only the simple example of an inhomo-geneous sheet which is a rectangle much broader than it is wide, with one portion of its width of resistance different from the rest. In the laboratory experiments this portion is an insulator. The quantity of principal interest is the horizontal velocity V of the sheet. Expressing this in dimensionless units, we have $(Pe) = Vh/\kappa$, which is a Peclet number, measuring the ratio of the advection of heat produced by the motion of the sheet to the rate of heat transfer by thermal conduction.

Experiments were performed in a plastic dish 50 cm × 50 cm filled with medicinal paraffin to a depth of 1–10 cm. The dish sat on an electrical heating element. The sheet most often used was a boat of 1·5-mm 'Perspex' of dimensions 4 cm × 8 cm with sides 5 mm high. The thermal resistance was a sheet of expanded plastic 3 mm thick.

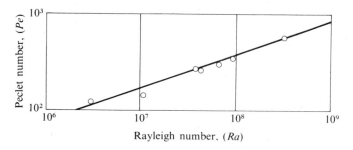

FIG. 11.3. Peclet number of a self-propelled sheet as a function of Rayleigh number; experimental results.

In brief, we find that if the Rayleigh number (Ra) is rather small the motion of the boat is erratic and does not have a preferred direction of motion. The Peclet number is rather small and the convection pattern underneath the boat is little different from that in the surrounding fluid. If, however, (Ra) is sufficiently large there is a quite pronounced persistent motion in a direction such that the high-thermal-resistance portion of the boat is at the rear. A typical run at $(Ra) \approx 10^8$ in a layer 5 cm deep gives velocities $V \approx 1 \text{ mm s}^{-1}$ and a corresponding Peclet number $(Pe) \approx 500$. The velocity V is not steady but fluctuates ± 20 per cent r.m.s. about its mean. Fluctuations of this amplitude are typical of high-Rayleigh-number convection.

At the higher values of the Rayleigh number we find, as with ordinary convection, that the gross properties of the flow are independent of h. This requires that $\{(Pe)/(Ra)\}^{\frac{1}{3}} = C$, which is of the order of 1, where C is a function of the parameters and in these experiments is about 1. The expression above is simply another way of saying that the velocity is independent of the fluid depth; rather it depends on the properties of the sublayer.

For the application of these results to crustal migration let us use the following estimates of orders of magnitude: $\gamma = 10^{-5} \text{ K}^{-1}$; $g = 10 \text{ m s}^{-2}$; $\Delta T = 10^3 \text{ K}$; $h = 10^3 \text{ km}$; $\kappa = 10^{-6} \text{ m}^2 \text{ s}^{-1}$; $\nu = 10^{16} \text{ m}^2 \text{ s}^{-1}$. The value $10^{16} \text{ m}^2 \text{ s}^{-1}$ corresponds to that at a depth of about 50 km. This is an appropriate depth for the present model, for we are most interested in the kinematic viscosity of the thermal sublayer on the base of the continent. In choosing $h = 10^3$ km I do not wish to imply that convection in the mantle is confined to this depth. Because we are dealing with a high Rayleigh number the results are independent of h. With the values above the Rayleigh number (Ra) is approximately 10^7. If we take the velocity V as $3 \times 10^{-10}\,\mathrm{m\,s^{-1}}$ (that is, $1\,\mathrm{cm\,yr^{-1}}$), the Peclet number (Pe) is 300; this is more or less what is found in the experiments.

Theoretical sketch: crustal self-propulsion

A very simple description of the process of crustal self-propulsion due to an imposed thermal anomaly is possible with the following model.

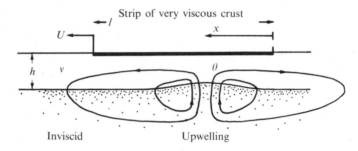

FIG. 11.4. Diagram for a simple model of self-propulsion.

1. We imagine a thermal anomaly, of excess temperature θ in the mantle. Because the ambient fluid is cool, the weight of a column of fluid above the anomaly will be less than that of a similar column in the surroundings. The anomalous region will thus be one of an upwelling of mantle material. As this flow reaches up towards the base of the crust it must turn and flow horizontally sideways and thereby exert a horizontal viscous stress on the base of the crust.

2. As before (p. 21), the horizontal outflow will be assumed to have its main effect in the more viscous upper mantle.

3. But here the horizontal pressure gradient driving the outflow arises from the horizontal differences of pressure which are created by the horizontal density gradient. Thus $p' \sim \gamma g h \theta / l$.

4. If the anomaly is centred a distance x from one edge of a crustal slab and the viscosity of the slab is much greater than that of the surrounding mantle, the net mean discharge velocity in the upper mantle away from the anomaly is $U \sim \{(1-x)\bar{u} - x\bar{u}\}/l = \bar{u}(1 - 2x/l)$, where $\bar{u} = p'h^2/3\mu$. Thus if the anomaly is centred beneath the slab, $x = l/2$, the flows to either

side are balanced, and no net motion of the slab ensues. But as the anomaly is displaced towards an edge the outflows are not in balance and a net motion occurs. The greatest self-propulsion velocity is $\bar{u} = \gamma g \theta h^3 / 3vl$.

Let us take an example with $h = 10^2$ km, $l = 2 \times 10^3$ km, and $\theta = 10^3$ K. Then $\bar{u} = 5$ cm yr^{-1}. This is quite a good representative value for relative crustal velocities.

Finally we need to note that temperature anomalies of order 10^3 K are quite possible. The thermal resistance from one part of the earth's crust to another can vary by about an order of magnitude so that anomalies of this order are possible at depths below about 30 km.

The interaction of the crustal slab with the mantle has produced a general circulation of mantle fluid beneath it in addition to whatever pre-existing flows there are. This circulation is of sufficient vigour to propel the slab relatively quickly over the mantle.

We have done rather better with this theoretical sketch than we should have expected. In this model we have in effect shown that the Peclet number is proportional to the Rayleigh number of the thermal anomaly. Strictly such a result is applicable only to flows with very small Peclet numbers, whereas here the Peclet number is actually large. To obtain a more detailed picture of what is happening we need to do a bit more.

Theoretical sketch: analysis of the general circulation

It is helpful to analyse the physical basis of the observed motions by considering a number of idealized numerical models.

Consider first the motion generated in an initially still layer of fluid of depth h at uniform temperature T_0 by a hot sheet at temperature $T_1 > T_0$. To draw attention here to the imposed temperature excess at the *top* of the fluid layer, I take the Rayleigh number based on the temperature excess as negative.

At Rayleigh numbers $-(Ra) < 10^3$ the problem is closely linear, so that $\psi \propto \{-(Ra)\}$ and the temperature field is unaffected by the flow—the Peclet number is small. For values of $-(Ra)$ from 10^3 to 10^4 the advection of heat by the flow becomes important and we find for $-(Ra) > 10^4$ that $\psi \propto \{-(Ra)\}^{\frac{1}{4}}$. We see at $(Ra) = -1$ a rather squat symmetrical cell with its centre below the edge of the sheet, but at $(Ra) = -10^4$ the cell is not at all symmetrical: it is considerably elongated, especially under the sheet, and its centre is displaced away from the edge.

Because the eddy is isolated near the edge of the sheet, the motion is independent of the width of the sheet provided that the sheet is wide. The contribution to the migration velocity from one edge is obtained by integrating the stress on the bottom of the sheet. When (Ra) is small,

$$(Pe)l = 0 \cdot 02 \{-(Ra)\}$$

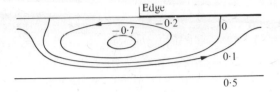

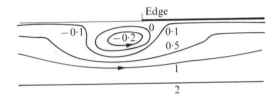

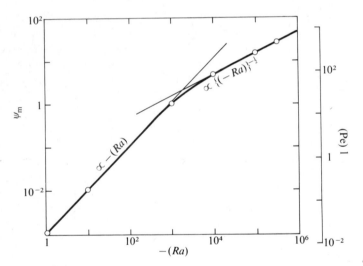

Fig. 11.5. Flow near the edge of a 'hot' crustal slab and the maximum transport as a function of Rayleigh number. Numerical simulation.

or otherwise

$$(Pe)l = 0.95\left\{-(Ra)\right\}^{\frac{1}{2}}.$$

We note in passing that the maximum shear stress occurs at the edge and has the value $2.2(Pe)l$. Thus we can construct the flow pattern relative to a migrating sheet by superimposing a horizontal velocity. These patterns (in which the observer is on the sheet) show a stagnation point ahead of the sheet in the vicinity of which ambient fluid is dragged down under the edge eddy. There is another stagnation point on the sheet.

One very important aspect of our problem appears here. I refer to the time required to establish the motion. At small Rayleigh numbers the scale $\tau \approx 0{\cdot}09$, which in dimensional units, taking $h = 10^3$ km and $\kappa = 10^{-6}$ m^2 s^{-1}, gives a time-scale of 3000 million years. Such evidence as there is, however, suggests that substantial movements are possible in 100 million years, so that we might guess that the scale is no more than 10^7 years—a factor of 300 less than the above figure. The models with low Rayleigh numbers can be discounted solely on these grounds. Beyond $(Ra) \approx -10^4$, however, the time-scale of such convective flows is approximately proportional to $(Ra)^{-\frac{2}{3}}$. A factor of 300 in time requires a factor of $(300)^{\frac{2}{3}} \approx 40$ in Rayleigh number. Thus, Rayleigh numbers at least of order 10^6 are indicated.

Rapid self-propulsion

To obtain a migration Peclet number of the order of 10^3 the simple edge model requires a Rayleigh number of the order of -10^6. Clearly if we are to consider such large Rayleigh numbers the layer of fluid itself will be in

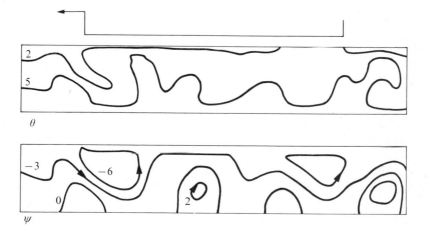

FIG. 11.6. Streamlines and isotherms in the upper mantle beneath a self-propelled sheet moving with uniform velocity. Numerical simulation.

vigorous convection regardless of the presence of the resistive sheet. We can therefore consider the consequent flow as a non-linear superposition of the free convective flow in the layer, together with flows induced by the sheet.

An example of such a flow has been simulated numerically. The sheet has width $l = 4$, is of effective thickness 1/15 of the layer depth, and has an imposed temperature of its base of 0·2 over the front three-quarters of its length and a temperature of 0·4 over the rear quarter. The flow field shows a horizontal flow, superimposed on which is a string of eddies. What is most striking about the flow is the intensification of the eddies near the edges of the sheet. Our sheet has both a forward and aft set of driving wheels.

In the numerical simulation the flow is seen relative to the continent where the migration velocity is at each stage of the calculation chosen so that the integral over the sheet of the viscous stress is zero. We see that the net effect of the eddies near the edges of the sheet is to accelerate the sheet, while the flow under the mid-portion of the sheet is to decelerate the sheet. The total effect on the sheet is zero: it is therefore moving with uniform velocity.

The typical values of the stress, which in dimensionless units are of the order of 10^3, are of some interest. We have the stress unit $\rho \nu \kappa / h^2 \approx 4 \times 10^{-4}$ bar, so that typical stresses at $(Ra) = 10^9$ are of order of 40 bar. These stresses are really quite small.

Consequences of self-propulsion

One of the notable criticisms of the old continental-drift hypothesis was that many of the large-scale active features of continents, which must involve effects in the upper mantle, remain fixed relative to the continent for periods of the order of 100 million years, while the continent is moving relative to the mantle. Such evidence is against a drift hypothesis, but it supports a propulsion hypothesis of continental motion because then the (thermal) effects in the mantle are related to the distribution of thermal resistance of the continent. Here the temperature anomalies in the mantle are carried along by the continent.

Our model has considered a rigid sheet. For the present purpose this is quite in order, especially in view of the fact that granitic rocks have a viscosity probably of order at least 10^2–10^3 that of ultrabasic rocks. If, however, we consider processes of time-scales greater than 100 million years, it is more appropriate to consider the sheet material also as viscous. We then have the intriguing possibility of a further interaction between the sheet and the upper mantle. Consider a localized shear-stress distribution on a portion of the base of the sheet which is tending to thicken it locally. If the viscosity of the sheet is rather small the hump of material will flow away into the surrounding areas nearly as quickly as it is produced. But if the viscosity is rather large the shape of the hump will closely follow the shear stress. Such a local thickening augments the local thermal resistance and thus alters the local convection in the upper mantle. In such an event there would be an intimate connection between large-scale tectonic processes and migration.

The essential feature of our model is the variation of temperature over the base of the sheet. As has already been noted (our definition of thermal resistance includes the effect of internal heat sources), this could arise not only from variations of thickness or thermal conductivity but also from variations of radiogenic heating.

The structure of the model mantle suggests the following.

1. The cold tongue descending under the leading edge of the sheet margin is not only cold but a region in which there are large velocity gradients across the tongue. In other words, this is a region of high strain rate in a more viscous material. This can therefore be identified as a possible earthquake region, the so-called Benioff zone. The vertical extent to this region is defined by the flow structure attached to the migrating sheet. In a model of this kind the depth of the 'earthquake region' is determined,

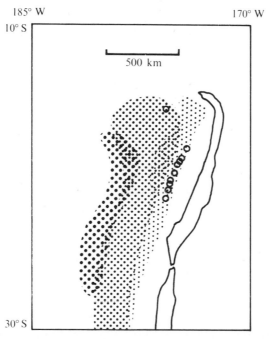

FIG. 11.7. A crustal sump: part of the structure in the Tonga–Kermadec region. The eastern crust is plunging westward beneath the western crust. Contour line encloses an area of the sea floor over 6 km deep. O, volcano; shaded area indicates location of earthquake epicentres of various depths: light, 100–200 km; medium, 200–500 km; heavy, below 500 km.

not by a change in the chemical structure of the mantle or by a dramatic change in relaxation time with depth, but by dynamical processes which have their origin in the crust and uppermost parts of the mantle.

2. The model does not concern itself directly with the oceanic crust; in effect we say that the oceanic crust is a thin superficial layer on top of the oceanic mantle. Nevertheless it is unlikely that the thin light sedimentary oceanic layer, about 0·5 km thick, can be buried in the mantle beneath the advancing continental margin. Rather, there must be a zone of accumulation, and once this zone is sufficiently thick and thereby hot

enough it can be expected to develop as a nearly independent subsystem (more about this in Chapter 12).

3. There is strong upwelling behind the leading edge, and if this is sufficiently intense volcanism may develop. But beneath the very viscous crustal material there may be little surface manifestation. This is, however, the sort of structure that has prompted the more detailed investigation below.

4. If there is little release of heat by volcanism across the sheet (as here) an accumulation of heat is available beyond the trailing margin that could lead to extensive oceanic rises and rather intense oceanic volcanism.

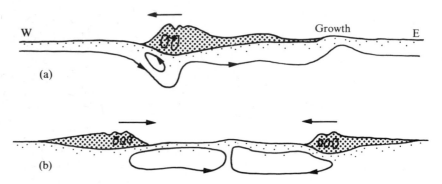

Fig. 11.8. (a) Diagram for an isolated migrating sheet with attached, trailing oceanic crust. (b) Diagram for a pair of opposed migrating sheets.

The model we have discussed so far has had a single area of anomalous thermal resistance. In nature the areal distribution of thermal resistance will be rather variable, as will the consequent distribution of horizontal viscous stress. Provided these stresses persist for a sufficient time and the regions of large stress gradient are regions of hot upwelling, differential movement of portions of the sheet will be possible. This is the topic of the following section.

If we accept the evidence that ocean-floor spreading rates agree with the rates of continental migration so that (some) continents are fixed to a sheet which includes fresh oceanic mantle, we have a situation, for example, like the region of South America and the western south Atlantic. The Andes region with its vigorous plutonism is one of low thermal resistance so that migration is relatively westward. The region of upwelling beyond the trailing edge is identified as the 'mid-oceanic' ridge.

Here the growth of the trailing edge and the sheet to the east (regarded in this context as a fixed reference) occurs to the west of the fixed margin. Throughout this process the oceanic thermal anomaly remains associated with the migrating sheet near the trailing edge.

As fresh oceanic mantle is 'welded' on to the trailing margin of the sheet some of the heat of this fresh material will be lost by volcanism. Provided the heat loss is sufficient the viscosity will be low enough for the fresh material to behave as an integral part of the migrating sheet.

In this situation the sheet to the east is also growing westward by the accumulation of fresh material. Also, in relation to the eastern sheet, the 'mid-oceanic' ridge is migrating westward.

So far we have considered an isolated mobile sheet. For a rise upstream of a migrating sheet, such as the eastern Pacific rise, we need to consider the interaction of two opposed self-propelled sheets. Here the intervening mantle does not have continental material in it and the presence of the upwelling is a response to the general circulation induced by the ingestion of oceanic mantle.

The view taken here is that 'mid-oceanic' upwelling is not the prime cause of ocean-floor spreading or continental drift. Rather it is seen as a direct consequence of the self-propulsion of sheets of continental type.

It is therefore a little easier to account for the pronounced offsets characteristic of 'mid-oceanic' ridges. If the prime cause is a mechanism for heating and consequent upwelling beneath ridges it is difficult to account for the persistence of ridges and for their offsets. But if the ridges are an induced feature, relatively small inhomogeneities of the lithosphere, especially viscosity variations produced by small temperature variations, can lead to local variations of relative movement. This topic is discussed more explicitly below.

Structure in continental areas

The modelling of the viscosity variation so far—a very viscous crust and a uniform mantle—is rather crude if we are to describe in detail the structures formed. In a more elaborate attempt to study the role of the viscosity

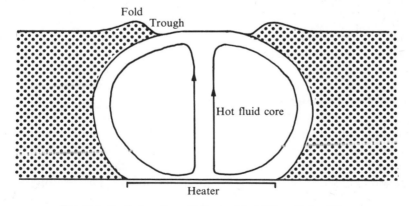

Fold
Trough

Hot fluid core

Heater

FIG. 11.9. Vertical section of jelly model of rifting; $(Ra) = 10^8$.

variation I have done some experiments in which the working material is jelly-gelatine dissolved in hot water and left to set together with a very thin layer of very viscous material on the top free surface (about 0·1 mm of an oil-based paint left to set).

Consider the case in which the system is cold and at rest and the heater is suddenly turned on. A bulbous 'fluid' region attached to the heater

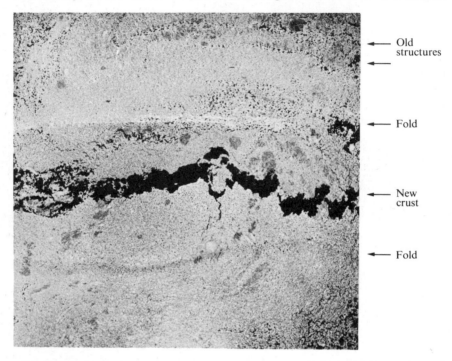

Old structures

Fold

New crust

Fold

FIG. 11.10. Photograph of jelly model of a rift system showing position of folds and internal structure.

gradually grows upward. The method of growth is quite different from that in a homogeneous material, where a buoyant element readily displaces adjacent fluid of the same viscosity. Here the adjacent fluid is so viscous that it can only be displaced by first lowering its viscosity by the convective heating from the 'fluid' region. (This must not, however, be confused with melting in which in addition the latent heat of melting must be supplied.) If the heater remains on long enough, the bulbous region ultimately penetrates the upper surface. If the Rayleigh number (which depends on the depth of the heater and the viscosity of the hot jelly) is small the subsequent gradual horizontal growth of the bulbous region is undramatic. On the other hand, if (Ra) is sufficiently large, for example, 10^8 in these experiments, a number of strong effects are seen in the surface region.

1. A fold-and-trough structure develops near the margin of the 'fluid core'. Further, if the heater is turned off, after all the jelly is again cold, the fold and trough remain. If then the heater is turned on again a new set of folds develops. In this way we can produce a succession of folds and troughs of different ages. The presence of the very viscous skin aids the formation of the folds.

2. In the region above the upwelling, pronounced surface fracturing of the very viscous skin occurs. This characteristically has the form of more-or-less linear features occasionally offset and with occasional transverse fractures.

3. In the 'fluid' region immediately beneath the surface and beneath the fold–trough region cold blobs are seen plunging downward and at the same time being swept along by the mean flow.

We can identify the fold–trough region with similar structures across some African rifts, margins of small basins such as the Irish Sea, and, on a large scale, with structures in Eastern Greenland. In the model the tensile stresses in the fold region are clearly insufficient to rupture the material, but in actual systems we would expect intense rupturing and vigorous intrusion in the folded region. Thus we see a possible relationship between dyke-swarms in the margins of a rifting region and the dynamical processes at depth.

Structure in sub-oceanic mantle

We have seen how the interaction between the continental mass and the mantle has led to large-scale circulations beneath the continent and substantial modification of the structure near its margins. But these induced circulations will also affect the structure in oceanic areas. The roughly uniform rates of sea-floor spreading indicate the presence of large-scale general circulations in the mantle. Such motions pose two general problems: the origin of the circulations, which we have looked at, and the manner in which they will affect the smaller-scale structures of the upper mantle thermal sublayer. This latter problem can be studied nicely in the laboratory. But to help our understanding of the structures produced in the more general situation we first look at the consequences of a simple uniform flow imposed on the region of interest.

Consider, therefore, the two-dimensional motion produced in a box through which there is a forced flow. The case with Peclet number $(Pe) = 0$ has already been referred to in Chapter 8. The intermittency and growth of buoyant plumes below the cooled surface are notable features. Here we concentrate on the case of non-zero Peclet number. If (Pe) is sufficiently small, the buoyant elements readily cross the box before being diffused

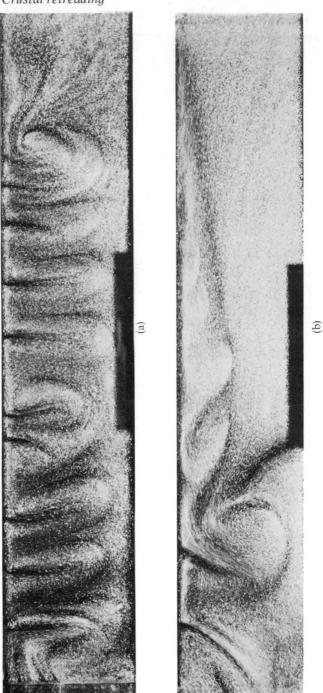

(a)

(b)

Fig. 11.11. Visualization of flow beneath a migrating sheet: (a) $(P_e) \approx 10$; (b) $(P_e) \approx 100$. The sheet is moving from left to right. (The rectangular shadow is of part of the apparatus.)

away. The production of buoyant blobs is less pronounced than in the case where $(Pe) = 0$, presumably since the motions in the eddy train are sufficient to remove the thermal energy from the sublayer. If (Pe) is sufficiently large, the flow is no longer everywhere dominated by the free convective motion. Three distinct regions can be seen: entry, intermediate, and mature regions. In the entry region there is a roughly regular production of buoyant plumes. Fluid particles near the upper surface are cooled as they are swept along until the sublayer locally becomes unstable. Each plume locally denudes the sublayer. The plumes are leaning over because the pressure is reduced in the warm region, and the horizontal velocity is therefore augmented at higher levels. The blobs pass though a period of maximum intensity and then diffuse away. Buoyant blob production is no longer so necessary in the intermediate region, where the relics of the plumes remain, and already the mean profiles have been considerably altered. Nevertheless, these motions are decaying, and in the mature region all effects of the entry region are lost and plume production recommences. Now the whole depth of the layer is penetrated by the blobs.

In brief, we have discovered that the simple mixing process of a layer stirred by the vertical movement of blobs randomly produced in the sublayer is strongly modified by the presence of a persistent large-scale circulation. If the Peclet number of this large-scale motion is sufficiently large, blob production in a system of finite horizontal extent is largely destroyed: the heat previously carried by blob production can be carried in the large-scale motion. But the large-scale motion so far considered, a uniform horizontal flow, is rather idealized, and we now search for a more realistic flow and a natural way of producing it.

Consider a long trough of viscous fluid heated on one vertical wall and cooled on the other, insulated below and cooling at the free surface, and in which the restriction to two-dimensional motion is removed by making the width greater than the fluid depth. The hot vertical wall is thought of as corresponding to the heat source below a region of 'mid-oceanic' upwelling.

For sufficiently small values of Rayleigh number (Ra) the only motion is a uni-cell filling the vertical cross-section. Viewed from above we see that the flow in the surface layer is uniform, as if the surface layer were a rigid sheet, with all fluid particles moving with the same velocity. For sufficiently large values of (Ra) we observe that relative motions develop in the surface. In a region, bounded above by the free surface, beyond a distance x_c from the upwelling flow arising from the hot wall, longitudinal rolls develop. In plan (a) (Fig. 11.13), the particles used to mark the fluid develop a striated pattern with a dominant transverse wavelength. In end elevation (b), we see that the striations are caused by eddies of alternating circulation confined to an upper portion of the fluid. In side elevation (c), the secondary motion is seen to grow downwards as it is swept along by the outgoing mean flow.

The mechanism leading to the onset of the secondary motion is made clear by considering the development of the temperature profiles in the surface region downstream from the hot upwelling. The upwelling forces a tongue of hot fluid underneath the surface. Cooling by conduction follows with the effect of broadening and diminishing the tongue. The depth of the peak will increase approximately as $x^{\frac{1}{2}}$, and the peak temperature as $x^{-\frac{1}{2}}$.

The upper region is one of unstable density gradient and the Rayleigh number of this unstable region will increase as x. Ultimately it will exceed a critical value, of the order of 10^3, when overturning will commence.

The pattern of volcanism to be expected from this structure, in addition to the major ridge feature of a central uplifted hot region which has been built into the present model, will be features transverse to the ridge axis of an

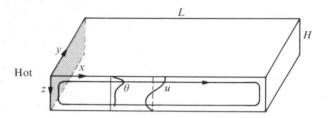

FIG. 11.12. Arrangement of laboratory model for study of sub-oceanic mantle structure.

alternating pattern of uplifted hot and depressed cold regions. Noting the model pattern, which has a tendency for breaks to occur along the longitudinal rolls, and the irregular nature of the blob production, it would be expected that the pattern would have occasional breaks and a spotty distribution of volcanism.

The data of topography and heat flow and magnetic anomalies are rather uneven, but the very low heat flows do correspond to depressions and the highs are predominantly near the flanking ridges. There are transverse features in the magnetic strips, presumably produced by a partial or complete loss of remanent magnetism in the hotter upwelling regions.

The horizontal scale of these structures is of the order of 100 km. We therefore have here a direct measurement of the scale of the sublayer of the earth's mantle.

All this suggests that, whereas we might have assumed that oceanic volcanism was randomly distributed, we now see that there is a distinct tendency for a correlation, not only with the main ridge, but the numerous transverse ridges. It is also now clear why there will be a progressive, though perhaps small, change in magmatic rock-type away from the ridge, since a substantial portion of the sublayer material is trapped in the layer as it is swept along. Occasional loss will occur through volcanism and a progressive

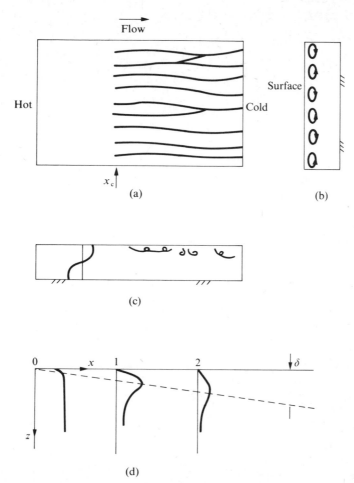

FIG. 11.13. Sub-oceanic mantle flow structure from a laboratory model. Sketch of the form of the developing secondary motion; (a) plan; (b) end elevation; (c) side elevation; (d) schematic temperature profiles at various distances from the upwelling.

ingestion of deeper mantle material as the layer grows in its outflow from the main ridge. Finally, this model suggests that oceanic volcanism may occur at a particular time at any distance from the main ridge. The model is thereby more in line with the evidence on ages of oceanic islands than the suggestion made some time ago that there was a pronounced increase of age away from the ridge.

12. Second-hand rock

IN NATURE there is no real waste; the end-product of one process is just the input to another. On the geological scene this is nowhere better demonstrated by the reprocessing that takes place in those sedimentary junk-yards, orogenic systems. Patiently collected over periods of the order of 100 million years, these worn-out sediments are transformed again in great spasms of activity into magnificent rocks.

The topographic clue
One of the most powerful global mass-transport processes is erosion and the subsequent dispersal of rock-substance as dissolved and suspended load in

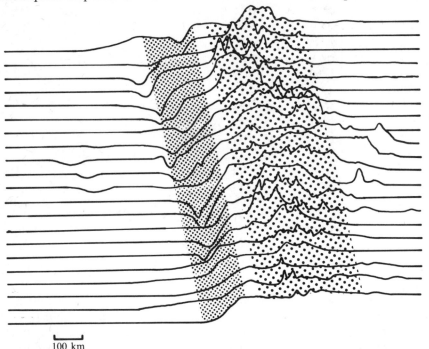

100 km

FIG. 12.1. Crustal dump: a slice of the Andes. Transverse profiles from 2° S, at top of diagram, to 46° S, at bottom of diagram, drawn every 250 km along the axis of the Andes. Transverse scale shown; vertical exaggeration 20:1; longitudinal scale 0·2:1. On the oceanic side (left) all profiles are taken to the mean depth, 4·5 km; on the continental side (right) to the mean land elevation, 0·2 km; all profiles lined up at sea-level. The active region (shaded) is 500 km wide and 5000 km long. Within it, 10^8 km^3 of rock-substance is being reprocessed. It is advancing from east (right) to west (left), taking about 20 million years to pass by. (Compiled from *Times Atlas of the World*, plates 119–21.)

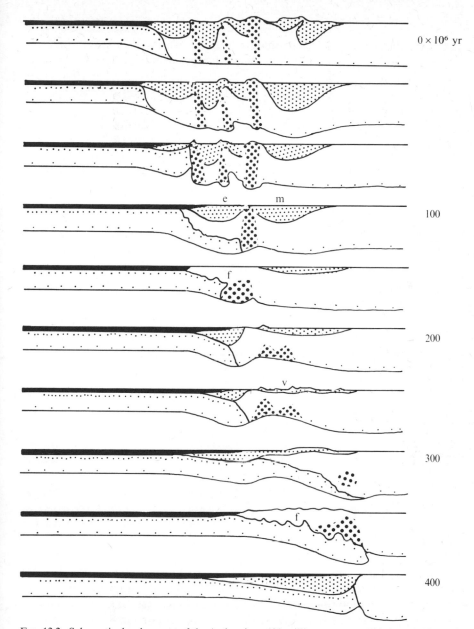

0×10^6 yr

100

200

300

400

Fig. 12.2. Schematic development of the Andes since 400 million years B.P. This is a kinematic summary based directly on field evidence. Labels show ages in 100 million years B.P. Two episodes can be seen: (*i*) an apparently mild one from before 400 million to about 250 million years B.P.; (*ii*) a vigorous one from about 250 million years B.P. to now. The diagrams represent a transverse section through the crust of horizontal extent 1000 km as: heavy line, left (west), ocean; below that, basaltic crust; plain areas, right (east), granitic crust; fine shading, sediments; medium shading, granitic plutons; coarse shading, partially melted granitic crust. Folding, metamorphism, and volcanism are more or less continuously active but notable occurrences are labelled: f, folding; v, volcanism; at 100 million years B.P. the so-called eugeosyncline, e, and miogeosyncline, m. (After C. J. Campbell.)

streams and oceans. Were it not for internal processes regurgitating the crust it would be worn down to near sea-level in a time much less than 100 million years. The rate of erosion varies enormously, but the global rate is such that a mass equal to that of all the land above 200 m is transported every 20 million years.

Because erosion and isostatic readjustment are so rapid the over-all topography of a region gives direct evidence of the presence of underlying recent activity, and in particular some idea of the density. For example, in an elevated region the over-all underlying densities must be lower than those in the surrounding region.

Global rate of sediment production

If we have a reprocessing facility it is necessary to be assured of adequate supplies. An estimate of the rate of supply of the sedimentary raw material can be made in two distinct ways: directly, by measurement today of for example suspended and dissolved matter in rivers, and indirectly by measurement of the thicknesses of sedimentary rocks.

The *gross* removal of matter from the land can be crudely analysed into two processes: chemical and mechanical. On the continental scale, chemical removal now at the rate $30 \, \mathrm{ton \, km^{-2} \, yr^{-1}}$ is independent of the mean continental elevation. The rate of mechanical removal per unit area increases with mean continental elevation at a nearly linear rate, so that we have the unexpected result that the rate is proportional to volume of the land mass above sea level, being now about $100 \, \mathrm{ton \, km^{-3} \, yr^{-1}}$—with a corresponding time-scale of order $30 \times 10^6 \, \mathrm{yr \, km^{-3}}$. Globally, the ratio of chemical to mechanical rates, with the above figures, is close to $1:3$. In the subsequent discussion the chemical rate will be ignored, its incorporation does not alter the argument.

If the amount of land mass has changed slowly throughout geological time we might then expect the over-all thicknesses of sedimentary rocks to be proportional to time. This is not so. The older the sedimentary rock the less there is of it. Of course, erosion acts on the sediments too. If we make the simplest assumption that the amount of a sedimentary deposit that is removed per unit time is proportional to how much remains, and that the total rate of erosion is constant, then both the amount of a particular deposit of age t and the total thickness of all deposits of ages zero to t is proportional to $\exp(-t/\tau)$ where τ is a time-scale inversely proportional to the erosion rate. Various estimates of the thickness of the so-called geological column have been made. If we use data on the cumulative thickness of the maximum known sedimentary deposits as a function of age we find the best fit to be: $\tau = 580 \times 10^6 \, \mathrm{yr} \pm 3$ per cent r.m.s.; total column thickness $= 210 \, \mathrm{km}$. The goodness of the fit suggests that the sediment production rate has been roughly constant throughout the bulk of geological time.

Theoretical sketch: the global time-scale of sediment production

We have obtained estimates for two global time scales: that of erosion, $T = 30 \times 10^6$ yr; that of total sedimentary rock, $\tau = 580 \times 10^6$ yr which are different by an order of magnitude. At first sight this is very puzzling, but it cannot be dismissed on the grounds that the estimates are crude. Consider the following very simple model:

(1) all the land mass is taken to be in a single continental block of thickness H, of density $(\rho - \Delta\rho)$, immersed in a mantle of density ρ;
(2) the depth d of ocean is constant;
(3) the mean land elevation h is determined by an erosion rate proportional to the volume of land above sea-level and the condition that the continental block as a whole is in isostatic equilibrium.

Thus from isostasy: $H\Delta\rho = \rho h + (\rho - \rho_w)d$, where ρ_w is the density of sea-water; and from erosion $dH/dt = -h/T$, so that $h = h_0 \exp(-t/\tau)$ where h_0 is the initial elevation and $\tau = (\rho/\Delta\rho)T$. Since $\rho/\Delta\rho$ is of order 10 we see that the two time-scales are compatible.

The global time-scale of sediment production is hence determined by the balance of removal by erosion and the elevation of the land owing to the consequent change in its buoyancy. This makes one think of those plate dispensers sometimes seen in a cafe where plates are loaded into a vertical cylinder in the base of which is a big spring so that as one plate is removed another pops up.

This is not the end of the story. Were it not for the relatively more rapid areal rearrangement of the crust as a whole, with its time-scale of order 10^8 yr, we would have an embarrassing excess of sediment. Since oceanic basins are so rapidly destroyed there is not the time for them to fill up with sediment. It is on these grounds that our assumption of constancy of ocean depth in this model is justified. Further we can obtain directly a global estimate of the average age of the oceanic basins. If we identify the so-called layers 1 and 2 of the oceanic crust as of sedimentary origin and estimate their thickness from measurements below the abyssal plains of the oceanic floor to be about 0·5 km and 1·7 km, we have a total ζ of about 2·2 km. Thus the mean age of the ocean floor is T (oceanic area/land area) (ζ/h), about 150 million years.

Nevertheless all that vast amount of sediment does have to be reprocessed.

Sweeping up the sediments

Numerous orogens have been found throughout geological time. Today, broadly speaking, we have two great orogenic zones, one ringing the Pacific, the other stretching across southern Asia and Europe.

An orogenic system may contain numerous elevated mountain ranges, predominantly on the sites of former geosynclines, and intervening depressions, often reflecting former basement ridges. The individual structures tend to follow the over-all trend but often the arrangement is higgeldy-piggeldy.

There are enormous variations in sediment thickness. Sediments will be accumulated in troughs:

(1) from local borderlands by transverse dumping;
(2) by longitudinal filling by coastal currents from distant areas;
(3) or scooped up off the ocean floor during crustal migration.

TABLE 12.1
Variation of river erosion

	Average sediment load per year of drainage area (ton km^{-2} yr^{-1})	Equivalent erosion (yr km^{-1})
Hwang Ho	2600	0.4×10^6
Ganges	1400	0.7×10^6
Colorado	380	2.6×10^6
Mississippi	97	$10 \ \times 10^6$
Amazon	60	$17 \ \times 10^6$
Congo	16	$62 \ \times 10^6$
Yenisei	4	$250 \ \times 10^6$

NOTE
The dissolved load carries an additional third.

TABLE 12.2
Sedimentary thicknesses in large basins

Eastern North America	5–9 km	Local depressions above continental crust on the shelf and off shore
East Africa	15 km	Marine facies
Gulf of Mexico	10 km	On oceanic crust
Caledonides of Britain	25 km	Currently identified
Urals	8 km	Little crustal shortening
Caucasus	10 km	Little crustal shortening
Donbass	10 km	Downwarped trough of the Pre-Cambrian shield
Andes	15 km	Cretaceous eugeosyncline

Consider an ocean floor of width 3000 km on which sediments have been accumulating at 1 cm every 1000 years for 100 million years to produce a layer 1 km thick. If now a continental block ploughed across this area and all the sediments were piled up ahead of the leading edge of the block into a trough 100 km wide, they would be 30 km deep. Such a trough of material floating in the mantle would stand about 2 km above sea-level. Of course, while all this is going on sediment will continue to be deposited on the ocean floor, and as soon as the trough material rises above sea-level erosion will set in. Nevertheless the possibility of accumulating deep troughs of sediments, indeed the common occurrence of such troughs, should be no surprise.

Orogenic time-scales

An individual orogenic system may have a long and very complex history, but typically a single collection and processing phase takes 200–300 million years, with the bulk of the processing taking about 50 million years and the

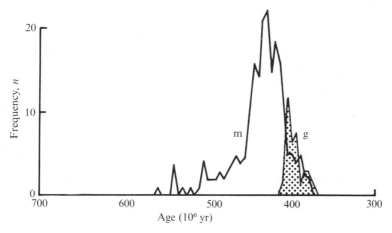

FIG. 12.3. Frequency distribution of K–Ar ages for rocks from the Caledonian orogeny of Scotland. m, metamorphic rocks; g, late granites. (After Fitch, Miller and Mitchell.)

appearance near the surface of granitic plutons signalling the end of a single convulsive processing phase.

Choice of process

If these sediments are to be strongly reprocessed there are two extreme possibilities.

1. Let us ask a simple but important question. Would it be possible, even if the pile were chemically homogeneous, for the temperature gradient alone to lead to overturning of the pile? With $\gamma = 2 \times 10^{-5} \, \text{K}^{-1}$, $v = 10^{16}$ $\text{m}^2 \, \text{s}^{-1}$, $\kappa = 0.5 \times 10^{-6} \, \text{m}^2 \, \text{s}^{-1}$, $h = 30 \, \text{km}$, and $\theta = 10^3 \, \text{K}$, the Rayleigh number $(Ra) = \gamma g h^3 \theta / \kappa v = 1.2 \times 10^3$. This is more-or-less the critical value for the onset of convective overturning. If the pile were somewhat thicker, convective overturning would be readily possible. Since several mountain chains have roots down to 70 km, Rayleigh numbers of the order of 10^4 are possible.

   We are forced to abandon this idea as it stands. Reprocessing could happen by simple overturning of the solid mass under extreme circumstances, but with such low Rayleigh numbers the process would be extremely slow, nor does it account for the fluxing of the sediments to produce the granitic plutons.

2. We have, of course, been shutting our eyes to the somewhat more obvious possibility that melting can occur. This is possible once temperatures reach about 700 °C in the deep sediments and perhaps 1200 °C in the upper mantle itself beneath the sedimentary blanket.

Up to now our rock-substance has been considered as a single phase. In other words, it is all solid as in the mantle or, if fluid, all fluid as in the core. But now we need to consider systems in which some of the rock-substance is solid and some molten: a sort of soup of solid and liquid rock-substance. Such so-called 'lithothermal' systems are the subject of detailed discussion in the following chapters. It might therefore have been better to consider lithothermal systems before mountain-building were it not that, in general, lithothermal systems are generally localized objects whereas orogenesis is clearly a phenomenon related to crustal development on a global scale.

For the moment, however, we need consider only two gross aspects of a partially melted material. First, owing to the heat transport of the melt in moving readily through the ambient solid the effective heat transport is greatly increased. This is as if the thermal conductivity were increased. Second, the whole mass will be lubricated and mobilized, not only by the presence of the melt itself, but owing to the increased heat transport and consequent temperature rise, leading to a much lower viscosity of the solid rock-substance and also a lower density.

Let us look into the consequences of the first of these effects.

Theoretical sketch: cooking the sediments

Consider a sedimentary pile growing into the mantle. Assume that as the pile thickens it remains in isostatic equilibrium and that as its base moves downward the material below moves away, largely horizontally. At some depth, say 100 km, the mantle remains undisturbed by the presence of the growing pile. In particular, assume that the temperature and heat flux are given there. New sediment is added to the top of the pile, and it is cold. Then, given the temperature distribution in the original mantle and the thermal conductivity at each depth, we can calculate the temperature distribution at later times. If we are also given the melting temperature at atmospheric pressure and its increase with pressure we can determine when partial melting commences. In partially molten material we can replace the normal thermal conductivity by an augmented value.

Throughout the accumulation of the material in the growing trough, heat will be entering the sedimentary pile from the mantle below. If the trough grows sufficiently slowly the heat flux through it will be the same at all elevations and the temperature will increase more or less uniformly with depth. If, however, the trough grows sufficiently rapidly heat will have penetrated only into the deepest sediments of the pile, the superimposed sediments remaining cold.

Let us look at a typical case given by these numerical simulations.

For up to 40 million years the sediments accumulate, the newest sediments remaining fairly cold but the older sediments progressively heating up.

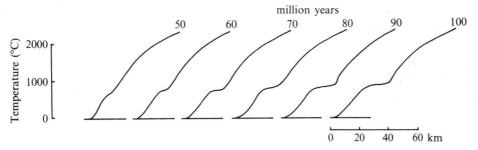

FIG. 12.4. Development of temperature in an accumulating sedimentary pile. Set of vertical profiles at 50–100 million years after lateral crushing of the pile commences, shown displaced sideways. Note the development of the zone of partial melting, indicated by the step in the profile, and the reduced temperatures below the zone.

At about 40 million years partial melting commences at the very bottom of the pile. To begin with its effect is unnoticeable.

After 60 million years the zone of partial melting is well established. Its effect, owing to the enhanced heat transfer, is to flatten the temperature profile. With extreme enhanced heat transfer, the variation of temperature with depth through the partially melted region will approach the melting-point–depth relation.

The temperature in the mantle immediately below the pile is increasingly lowered in response to the high heat-transfer rate in the partially melted sediments above, and extra heat is drawn from below. This zone of reduced temperature and high thermal gradient in the mantle grows downward into the mantle but has reached an equilibrium form after about 100 million years.

The effects of these changes in the rock-substance of the sedimentary pile will be, in the upper layers, to produce thermal metamorphism of grade increasing with depth until the melt-front is reached. Below this the fluxing of the ambient rock-substance by the melt will produce extreme metamorphism, perhaps granulites near the melt front but proto-granite or granodiorite well below the front.

The discussion has concentrated attention on the middle of the trough. On the margins the effects will be muted. For example, the shallower sedimentary layer may not become granitized at all. Further, we have naïvely assumed that all the sediments are uniformly and evenly dumped on the top of the pile. But for sediments being swept up by an advancing continental margin, for example, the situation is obviously much more complex. Sediments near the margin, having been collected earlier, will have relatively longer to bake.

TABLE 12.3

Cooking model data

Energy equation	$\dfrac{\partial \theta}{\partial t} = \dfrac{\partial}{\partial z}\left(\kappa \dfrac{\partial \theta}{\partial z}\right)$
Thermal diffusivity	(1) No melt, $\kappa = \kappa_* = \kappa_0\{1 + q(z)\}$
	(2) Melt, $\theta > \theta_m = \theta_0 + \alpha z$
	Where α is the melting-point gradient and θ_0 is the melting-point at $z = 0$
	$\kappa = \kappa_*\{1 + \beta(\theta - \theta_m)/\theta_m\}$
Structure	(1) Sediments, depth $= pt$
	(2) Basalt layer, 5 km thick
	(3) Mantle, undisturbed below 100 km
Sediments	$p = 4$ km per 10^6 yr
	$\theta_0 = 700\,^\circ$C
	$\alpha = 12\,^\circ$C kbar^{-1}, the same as wet granite
	$\beta = 100$
Basalt	$\theta_0 = 1200\,^\circ$C
	$\alpha = 6\,^\circ$C kbar^{-1}, dry
	$\beta = 100$
Mantle	$\theta_0 = 1400\,^\circ$C
	$\alpha = 3\,^\circ$C kbar^{-1}
	$\beta = 100$
Phenomenological parameter β	Guesswork; to give acceptable fit, a wide range is suitable
Variation of thermal diffusivity $q(z)$	Chosen such that for no sediments, $p = 0$, the steady-state temperature distribution $\theta(z)$ is that given in Chapter 8 with $\theta_0 = 2500$ K and surface temperature zero.

The convulsive phase of orogenesis

The gradual accumulation of the thick pile of sediments within the trough is sometimes followed by a rapid internal reorganization of the material during which relative vertical displacements of the order of 10 km occur over horizontal distances of the order of 10 km in times of the order of 1–10 million years, these effects persist for about 100 million years.

These gross structural rearrangements bring us to the second of the effects of partial melting, the increased mobility of the layer as a whole. Here we have a simple case of a gravitational instability. The partially molten material at depth will be less dense than the material above, and provided the density difference is sufficient (more about this in the next chapter), the

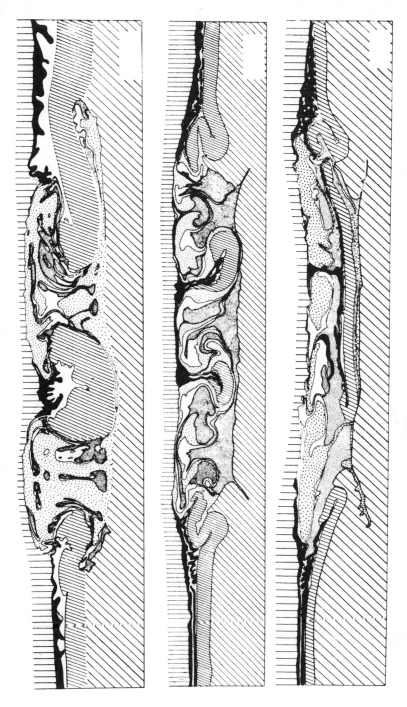

FIG. 12.5. Laboratory models of orogenic systems. Tracings from photographs of vertical sections. Equivalent time of development about 200 million years. (After Ramberg.)

pile as a whole will begin to overturn. Of course, once it has overturned it has shot its bolt, since the density contrast cannot be maintained.

Laboratory studies of these processes, initiated and dominated by vertically acting buoyancy forces, have been made. Layers of plastic materials, simulating the strata in the sedimentary trough, are placed together on the laboratory bench where the processes are slow and then speeded up in a centrifuge wherein the centrifugal force, simulating gravity, is typically $10^3 g$. During the period in the centrifuge the unstable mass distribution rearranges itself to a more stable state, in the course of which structures develop which are similar in shape and arrangement to those found in certain orogenic belts.

Associated basic volcanism

Before granitization has occurred any basaltic melt of density about $2 \cdot 7\,\mathrm{g\,cm^{-3}}$ produced either directly from the remaining oceanic crustal material or by differentiation from the mantle has to pass up through sediments of densities as low as $2 \cdot 4\,\mathrm{g\,cm^{-3}}$. For a given depth of the source of the basaltic melt there will be a thickness of sediments through which the basaltic melt cannot pass: it is too heavy. Thus, a rapid accumulation of a deep pile will prevent the surface manifestation of basic volcanism, except perhaps on the extreme margins of the trough.

Later, however, if a substantial part of the sediments has become granitized to densities of typically $2 \cdot 7\,\mathrm{g\,cm^{-3}}$, basaltic melts can pass relatively freely through the sedimentary pile: the fluid would have neutral buoyancy.

If the material below the sediments has not partially melted by the time that partial melting of the sediments has started, as here at 40 million years, the opportunity for mantle melting is lost because of the drop of mantle temperature. Melting will not then begin until the sedimentary accumulation ceases and erosion has removed some of the pile.

Crustal feedback

If the primeval crust were granitic and covered the globe beneath $2 \cdot 5\,\mathrm{km}$ of water, a suggestion made in Chapter 6, the mechanism of sweeping up the granitic crust would be the only method of creating sufficiently thick piles for orogenesis, since with no land erosion would not operate.

An interesting possibility can now be recognized: feedback from the changing conditions in the sedimentary pile at the leading edge of the slab to the over-all motion of the slab itself. Clearly the effective thermal resistance of the sedimentary pile will change throughout its development. At first, when the pile is thin, the extra thermal resistance will be negligible. As the pile thickens the thermal resistance will rise, and if a very thick pile develops its effect could be to slow down or even reverse the motion of the slab: remember, motion is away from the region of high thermal resistance.

But once either convective overturning or plutonism due to melting commences in the pile, its thermal resistance will drop to a very low value and the original slab motion will be restored and accelerated. This faster slab motion will then persist until the convulsive orogenic phase ends with the rapid erosion of the system.

13. Rock-mushrooms

AN OROGENIC system is a global-scale structure of partly reprocessed rock-substance produced merely by collecting a large enough pile of sediments. On a smaller scale, the scale of the mantle and crust, we find other structures, such as large igneous complexes, which have been produced by dynamically much more vigorous processes.

Clearly, there are three typical possibilities for the state of the rock-substance: molten, partially molten, or below its melting-point. For example, in the case of a partially molten system we apply the studies of convection in a permeable medium. From this point of view we can refer to a lithothermal system.

A lithothermal system is a heat-transfer mechanism within the earth that relies for its operation on the transport of rock-substance, though not necessarily the discharge of rock-substance at the earth's surface, and produces a region in which the heat flow is different from normal. The transported rock-substance need not be molten. We concentrate our attention on lithothermal systems in which a portion of rock-substance is molten and the rock matrix itself does not move. As with hydrothermal systems, we recognize three major elements in a lithothermal system: a heat source, the deep system, and the surface system. We assume the existence of a heat source at the base of the upper mantle, at a depth of some hundreds of kilometres. The deep system is the principal concern of this chapter. The surface system is that part of the lithothermal system which is strongly influenced by the presence of the surface. It is a zone up to about 10 km deep in which, for example, the manifest features of volcanism arise. The surface system is treated in the following chapter.

There is an obvious and important distinction between hydrothermal and lithothermal systems. Within the body of a hydrothermal system the working fluid, water, is conserved, whereas within a lithothermal system the working fluid, rock-melt, may not be conserved, simply because the melt may freeze. Also, the constitution of the melt in its percolation through the rock matrix will progressively change, especially in so far as it remains in thermodynamic equilibrium with the matrix. We see immediately the principal difficulty in analysing our problem, namely, that the permeability will vary with position as the melt fraction varies and further, the kinematic viscosity is a rapidly decreasing function of temperature.

Two essential classes of questions face us when we consider the phenomena of igneous intrusion. First, we have problems concerning the origin and mechanism of the process which allow the production of intrusions; second, there are problems of understanding how these mechanisms, and any others subsequently brought into play, determine the structure of intrusions. The scale of intrusions and the large quantities of heat required force us to the conclusion that these phenomena have their origin in the mantle.

Permeability: a continuum idealization

When a fluid permeates a porous matrix, in which the voids are connected, the actual path of individual fluid particles cannot be followed in detail— unless one had the facilities of the Archangel Gabriel—but the flow can be represented macroscopically by a relationship first discovered by a French sanitary engineer, M. Darcy, working in Dijon in 1856. He noted that the discharge from some filtration beds was proportional to the head of water across the bed. Thus idealizing the system as a continuum, if the mean mass flow is q per second across unit area under a pressure gradient dP/ds, where P is the pressure at the point s along a spatial-mean streamline:

$$q = -(k/v)\,dP/ds.$$

This relation involves the permeability k, a property of the matrix, or rather its voids, and the kinematic viscosity v, of the working fluid. For a matrix in which the porous structure is fairly uniform and has a well-defined length-scale Δ, such as the mean grain diameter for a bed of sand, $k/\Delta^2 \sim 10^{-4}$. The usual unit of permeability is based on quantities measured in the cgs system, except that viscosity is expressed in *centi*poises (cP) (the viscosity of water is about 1 cP) and pressure in atmospheres. Thus, 1 darcy $\approx 10^{-12}\,m^2$.

Similarly, the effective thermal diffusivity for convection in a permeable medium, $\kappa_m = K_m/\rho c$ where K_m is the thermal conductivity of the fluid-saturated matrix; ρ is the density and c the specific heat of the *fluid*.

In so far as lithothermal and hydrothermal systems are dominated by the permeability, the remainder of this book is a study of the thermally driven phenomena which can be understood in terms of the permeability concept.

Theoretical sketch: permeability of a given microstructure

Consider a matrix constructed from a regular cubic array of blocks of side Δ with narrow spaces of width $\hbar$ separating them. The permeability arises from the joint structure. For slow viscous flow we find the permeability $k = \hbar^3/6\Delta$ and the porosity $e = 2\hbar/\Delta$.

For example, in a small-scale structure with $\Delta = 1\,m$, $\hbar = 0.1\,mm$, we have $k = 0.2$ darcy. This is very much larger than the permeability of individual lumps of rock—typically of the order of 1 millidarcy—more than ample to run the systems discussed later. Relatively small amounts of jointing produce substantial permeability.

Free convection in a homogeneous permeable medium

The problem of discussing heat and mass transfer in a permeable medium can be treated in a similar way to that of a viscous fluid, as in Chapter 7. Thus, for a permeable volume of vertical extent h, across which there is a temperature difference ΔT, the flow is dominated by one dimensionless ratio, the Rayleigh number for a permeable medium $(Ra)_m = k\gamma g h \Delta T / \kappa_m v$.

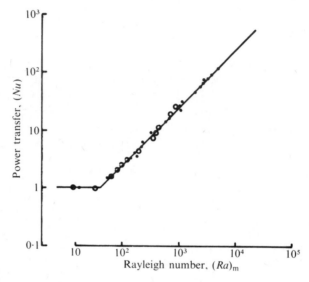

FIG. 13.1. Heat-transfer characteristic for free convection in a horizontal permeable slab. Dimensionless power transfer (Nu) as a function of porous medium Rayleigh number $(Ra)_m$. Note: $(Nu) = 1$ corresponds to zero convection. Experimental data indicated. Large natural systems operate in conditions corresponding to the upper part of the diagram.

Notice that this differs from the ordinary Rayleigh number mainly by h^3 being replaced by kh. As always, the prime quantity of interest is the power Q transmitted through the volume. Experimental data for a uniform horizontal slab heated over all its base show the following.

1. For $(Ra)_m \lesssim 40$ there is no convective motion. This is consistent with a prediction of the critical value $4\pi^2$.

2. For values of $(Ra)_m > 40$, the Nusselt number (Nu) closely follows: $(Nu) = 0 \cdot 025(Ra)_m$. In dimensional form this becomes

$$Q/(\text{heated area}) = k\rho c \gamma g (\Delta T)^2/40v,$$

a quadratic relation between Q and ΔT, and independent of h and K_m.

Experimental studies of two-dimensional flows in a permeable medium are conveniently done in a thin vertical cavity of thickness $a \ll h$. Hele Shaw showed in 1898 that flows in such a cavity were the same as in a medium of permeability $k = a^2/12$. Although the motion in a Hele Shaw cell is only two-dimensional, there is the great advantage that the flow can be visualized, for example, by means of suspended particles.

Intrusions

A common notion in volcanology is that of a magma reservoir. This is a convenient concept for discussing the behaviour of a lithothermal system over a short interval of time. The origin of the reservoir is, however, a more

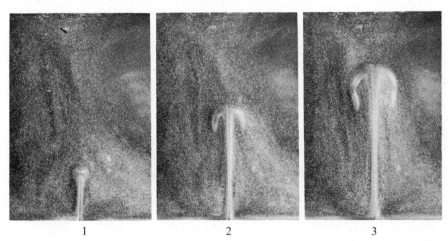

1	2	3

FIG. 13.2. Simplest model lithothermal system, an intrusion. Single blob produced from a small heat source turned on for a short time. Photographs at subsequent times 1, 2, and 3. Hele–Shaw cell. Depth of natural system typically up to 100 km

difficult question. The simplest idea that comes to mind is that of a rising hot blob, a system analogous to the rise of a salt dome.

Consider therefore the following simple model. A portion of the base of the upper mantle has its temperature raised for a time t_1: in other words, we have a local temperature fluctuation. After a time we observe a blob of hot fluid detach itself from the base of the layer and percolate upwards. For example, at $(Ra)_m = 10^3$ we find that the blob has completely traversed the layer by dimensionless time $t = 0.016$. Our model experiment is restricted to rather small values of $(Ra)_m$, so that there is a problem in extrapolating to the natural systems. For flows in a porous medium it is, however, a good first approximation to take time proportional to $1/(Ra)_m$ and velocities proportional to $(Ra)_m$. For example, imagine that such a blob originating at a depth of 100 km took 10^5 years to produce the maximum surface heat flux, corresponding to a vertical velocity of order 3 mm d^{-1}. Since the time-scale is 3×10^8 years, we need $(Ra)_m \approx 10^5$ and $k/\nu \approx 10^{-7}$. Assuming a

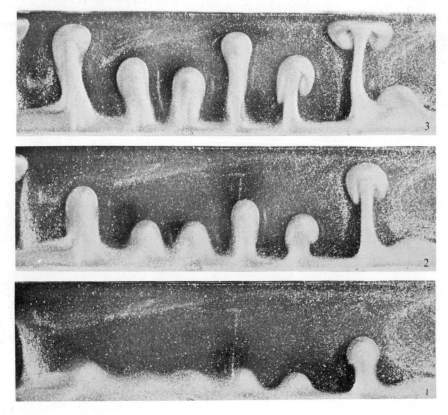

FIG. 13.3. Multiple blob production by uniform heating from below over a horizontally extensive area. Photographs at times 1, 2, and 3. Depth of natural system typically up to 100 km.

mean value for v of, say, $0 \cdot 1 \text{ m}^2 \text{ s}^{-1}$ this implies that the matrix through which the melt is percolating has a grain-size of the order of $0 \cdot 01$ m.

We observe that if t_1 is sufficiently small the blob never detaches itself from the lower surface. This is simply because if the heated region is small the motion is dominated by diffusion. We may estimate crudely a minimum (dimensionless) time t_1 by requiring that the Rayleigh number, based on the thickness of the heated region near the source, reaches the critical value of $4\pi^2$. Since the thickness of the heated layer is of the order of $2\sqrt{t}$, the minimum time is approximately $400/(Ra)_\text{m}^2$.

With $(Ra)_\text{m} \approx 10^5$, as given above, the minimum time is about 10 years. Hence in this case any fluctuations of temperature at the base of the lower mantle which persist for periods of about 10 years or more will lead to the production of hot blobs.

Accepting for the moment the figures given above, we see that the upper

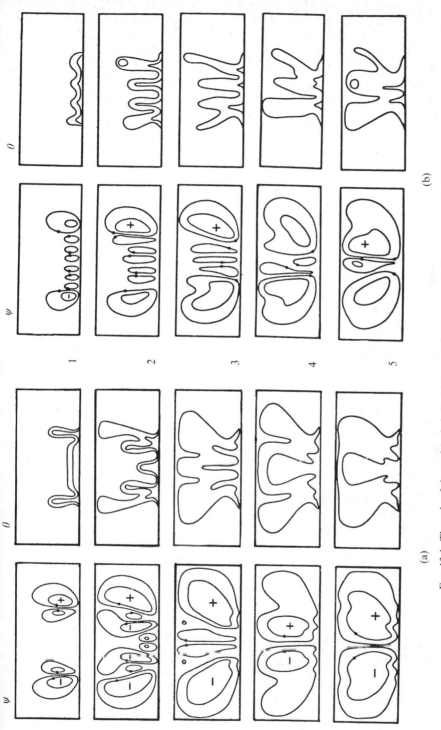

ψ 0 (a) ψ 0 (b)

Fig. 13.4. The role of thermal 'noise'. Development of free convection from rest at Rayleigh number $(Ra)_m = 400$: (a) without thermal noise; (b) with thermal noise. The time-intervals 1, 2, 3, 4, and 5 are in the time units $0\cdot01\ \mathrm{h}^2/\kappa$.

mantle is extremely unstable. The consequent motions will therefore be much more elaborate than those just described. Consider, then, a situation in which a uniform heat flux is produced at the base of the upper mantle over a localized region and in which at $t = 0$ the temperature is raised to $\Delta T(1 + \varepsilon)$, where $\varepsilon = \varepsilon(x)$ is a random function of position of range ± 0.25. The behaviour of this system shows the isotherms and streamlines at various times. When there is no noise the development of this system is regular and symmetrical. This is not the case when we have thermal noise. We then see four rising hot columns which, owing to the random initial temperature and the extreme instability of the system, grow at very different rates. The left-hand column has virtually disappeared by $t = 0.04$. The position of greatest surface heat flux changes considerably, being on the right-hand side at $t = 0.03$, above the second column at $t = 0.04$, and above the third column at $t = 0.05$. If the system is observed for a longer time it is seen gradually to become steady, with a single column rising above the middle of the heat source.

The consequences of the increase in permeability as the blob rises might be as follows. While the melt remains in thermodynamic equilibrium with the matrix, and this is likely to be the case except for the most intense penetration, the melt will tend to consist of the lower melting-point constituents of the matrix. For those elements which are more readily retained in the low melting-point fraction, considerable refining is possible. The upper part of the uprising blob, since it is at lower pressure, will begin to melt first, and this product will be the lowest-melting fraction of the system. This liquid will have lower density and will consequently begin to rise faster. The continuation of this process will result in a column of liquid which will be inhomogeneous, being least basic at the top and increasingly basic at the bottom. Hence the tacit assumption that magma reservoirs always contain initially homogeneous magmas which later differentiate is false. In any dynamical model of volcanism, such as that presented here, the magma reservoir will not be homogeneous; rather there will be nearly continuous variations within it. Since, however, there is a continuous circulation of magma as the reservoir rises we can anticipate that the distribution of the various fractions will be even more complex.

Volcanism

So far we have concentrated attention on isolated blobs of less viscous jam breaking through to the surface. If we take a broader view we notice that these blobs often occur in groups clustered in space and time. These volcanic complexes are a common feature of volcanism on the earth. For example, in a schematic geological map of such a system, the Taupo volcanic system, we see remnants of sheets of ignimbrite penetrated by rhyolite domes, which

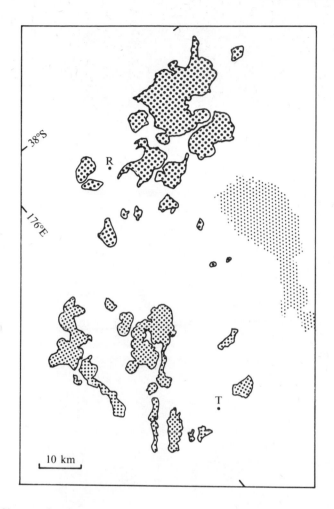

FIG. 13.5. Clusters of rock mushrooms. Plan view of the crust in the New Zealand volcanic zone, showing two groups of rhyolite intrusions, one near Rotorua (R), one near Taupo (T). The lightly shaded area in the east is an elevated portion of the graywacke basement. Otherwise the ground is covered with volcanic debris and ignimbrite flows. All these rocks are granitic. The intrusions shown are identified visually. There are a great many more that can be identified by the spatial variation of the magnetic field. Note the map is oriented nearly NE–SW.

themselves are associated in groups. There are two major groupings, in the upper and lower parts of the map, in the Taupo area.

An experiment in the laboratory shows the following.

1. A strong upwelling has already appeared in the north, and deep in the layer elsewhere numerous small blobs are beginning to rise.

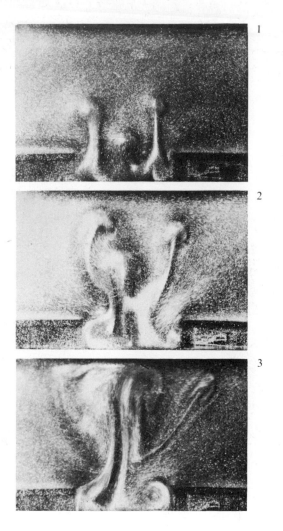

F_IG. 13.6. Vertical section of a model volcanic zone; temporal development. Photographs at times 1, 2, and 3.

2. One of these blobs is penetrating the surface.

3. The whole layer is in vigorous motion.

4. The convection is at its maximum development; some later blobs in the south have penetrated the horizontally outflowing material of earlier blobs.

The whole sequence, scaled to the Taupo system, represents a time-interval of the order of 10^6 yr.

In view of the viscosity variation we may not be content with a single-layer model and will prefer a two-layer model with a lighter more viscous layer on top. Here we see the results of temporarily heating the base of the

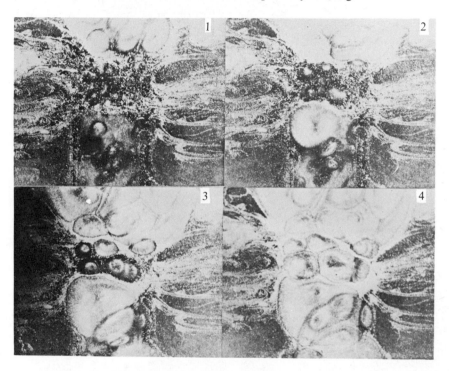

FIG. 13.7. Plan view of a model of a volcanic zone and its temporal development. Photographs at times 1, 2, 3, and 4.

lower fluid. We are looking down through the upper layer on to the top of the lower layer (which is seen as a rather pimply surface). Penetrating into the upper layer we see a pair of ring complexes and numerous small intrusions. All the motions revealed by the streaks are in the upper layer.

Structure from viscosity variations

So far I have considered only those structures that arise during penetrative convection and are dominated by the buoyancy forces. Further structure can arise through mechanisms which rely on non-homogeneity. One of these mechanisms involves the viscosity differences between the interior of an advancing penetrative 'front' and its surroundings. The mechanism is equally

FIG. 13.8. Laboratory model of a two-layer system. As in Figs 13.6 and 13.7 with a superimposed layer of extra depth 10 per cent. Mature stage of penetrative convection.

applicable to flow in a viscous fluid and in a porous medium but here I refer only to the latter case.

If the interface between two fluids 1 and 2 is made to move vertically upwards with velocity W because fluid 2 is being forced into the region occupied by fluid 1, the interface is unstable if

$$\left(\frac{\mu_2}{k_2}-\frac{\mu_1}{k_1}\right)W+(\rho_2-\rho_1)g<0,$$

where μ, ρ, k refer to viscosity and density of the melt and the permeability of the rock matrix. In particular, note that in the extreme case where $\rho_1 = \rho_2$, instability always occurs with $W > 0$ for $\mu_2 < \mu_1$, the less viscous

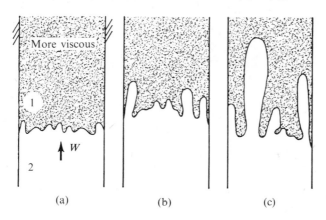

(a)   (b)   (c)

FIG. 13.9. Finger growth as a viscous fluid is forced into a more viscous fluid. Development in time (a), (b), (c). Tracings from photographs. (After Saffman and Taylor.)

fluid advancing into the other. This mechanism is easy to understand intuitively: the less viscous fluid flows more freely, and by creating its own channels needs to do less work displacing the other fluid. In laboratory experiments it is found that once instability occurs fingers of the generally less viscous lower fluid penetrate into the upper fluid, some fingers growing very much more rapidly than others.

The model I have in mind then is the following. Owing to a local instability of the upper mantle, penetrative convection commences. As one of the blobs rises and accelerates it may reach and exceed the critical velocity W—the fluid within the blob being hotter and thus less viscous—whereupon the upper 'surface' of the advancing blob breaks into numerous fingers, which themselves may break up further. Thus, from a single simple blob many fingers, that is, individual intrusions, can arise.

Structure from diffusivity variation

I now come to an intriguing phenomenon which could be possible only in a lithothermal system, a high proportion of which is melt. I first describe the laboratory studies with only two quantities which have different diffusivities. Salt and heat are often used with diffusivities of the order of $10^{-9}\,\mathrm{m^2\,s^{-1}}$

FIG. 13.10. Onset of multilayered convection in a thermosolute system heated from below. NOTE. (a) one layer already formed at the bottom and vigorously convecting; (b) sharp interface at top of first layer; (c) fluid being stripped from interface to form next layer; (d) undisturbed region.

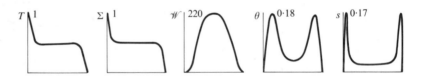

FIG. 13.11. Thermosolutal convection; profiles of horizontal means: temperature T; solute concentration Σ; and r.m.s. values of the fluctuating vertical velocity W, temperature θ, and solute concentration s. $(Ra) = 3 \times 10^5$, $\gamma = 2$, $\varkappa = 0.1$.

and $10^{-7}\,\mathrm{m^2\,s^{-1}}$ respectively, giving a diffusivity ratio $\varkappa = 10^{-2}$; but the use of other materials such as sugar and salt with $\varkappa = 0.3$ leads to the same phenomena. A layer of fluid is stabilized by a uniform vertical salinity gradient and is then heated from below. Let us write $\gamma = a_S\,\Delta S/a_T\,\Delta T$, where the fluid density is $\rho = \rho\,(1 - a_T\,T + a_S\,S)$, ΔS, ΔT are the salinity and temperature differences applied across the layer, and a_S and a_T are coefficients independent of S and T. Clearly γ is a measure of the static stability of the system. If $\gamma < 1$, the temperature is dominant and ordinary convection (complete overturning of the layer) will ensue; if $\gamma = 1$, the system is in balance; and if $\gamma > 1$, we might expect the system to be stable. It is one of the most interesting convective phenomena discovered in recent years that even if $\gamma \gg 1$, so that the system is gravitationally statically very stable, vigorous, turbulent convection can ensue.

If a layer with $\gamma \gg 1$ is suddenly heated from below, for example, the following sequence of events occurs. After a time, convection commences in a thin band at the bottom of the layer. A sharp interface divides this vigorous flow from the fluid above. After a further interval a second convecting band appears on top of the first. This process continues until there are many horizontal bands within the fluid. Structures of this kind are well known in the ocean.

Consider the structure of a single layer. The following features should be noted.

1. The mean temperature field T has thin boundary layers on the wall and a nearly uniform interior, as in ordinary convection.

2. The mean salinity field Σ has a similar structure, except for very much thinner boundary layers.

3. The fluctuating temperature and salinity fields θ, s have a double peak with the peaks near the margin of the boundary layers. Many of these features are similar to those of ordinary convection; but there is one feature peculiar to the thermohaline case. Since the thermal boundary layer is thicker than the salinity boundary layer (approximately in the ratio $1/\varkappa^{\frac{1}{2}}$), there is a region of positive buoyancy in the outer part of the thermal boundary layer. The interior uniformity could not be maintained were it not for some vigorous process at the margin of the boundary layers, within which diffusion is the dominant process. An approximate indication of the strength of this outer region is given by the thermal boundary-layer Rayleigh number, which for this example is about 6×10^3. This indicates vigorous motion, since in ordinary convection the corresponding figure is about 10^3.

There are thus three main processes in the flow: diffusion out of the boundary layers; a vigorous mixing region at the margin of these layers; and a well-stirred interior. But the essential ingredient in the maintenance of these processes is clearly that the diffusivity ratio $\varkappa \ll 1$, so that even if the total stability effect of the salinity field is large, namely, $\gamma \gg 1$, the drop of salinity is confined to boundary layers which are thinner than the thermal boundary layers.

I suggest that some banded layering in igneous intrusions, such as that found in certain ultrabasic sills and in the notable case of the Ilimaussaq intrusion in Greenland, is produced by this process. The simplest case would be that in which at some time in its life the intrusive material was nearly all melt which was being progressively heated from below. The role of the salt is replaced by some dominant constituent which will form the bulk of the ultimate rock-matrix. Actually there will be several constituents and several

FIG. 13.12. The layered Ilimaussaq alkaline intrusion of South Greenland. (Photographs by courtesy of B. L. Nielsen, Greenland Geological Survey.)

diffusivities. Layers will form successively, one above the other. Ultimately the intrusive body will begin to cool owing to termination of the heat flow from below and cooling from above. Crystals will form and fall but will often be remelted in the hotter layers at depth. We thus have a very powerful refining mechanism with material, crystalline and liquid, largely trapped in individual layers but with a small progressive downward flux of matter in crystalline form, and an upward flux of liquid due to transport across interfaces. Volatile material will easily move to the upper parts of the intrusion and may be responsible for a large part of the heat loss. These volatiles may be responsible for the ore bodies associated with this class of intrusion.

14. Magma clocks

SOME ROCK is reprocessed more or less on the spot during orogenesis but a far larger quantity of brand-new rock is spewed out of the upper mantle to

FIG. 14.1. Aso-san, Kyushu, Japan one of the earth's largest calderas, about 20 km across, looking east over the central cones active since A.D. 1238.

plate the oceanic crust and parts of the continental crust. How do these volcanic rocks actually get out of the interior?

Lithothermal system: surface zone

In a hydrothermal system the surface zone is the region in the vicinity of the water-table where direct flow, evaporation, and flashing can occur. There is an analogous zone for a lithothermal system in which degassing can occur. There is, however, an additional feature of the surface zone of a lithothermal system which, apart from relatively rare hydrothermal eruptions, is distinctive. Pressures in a lithothermal system are quite large enough to exceed the breaking strength of rock. Thus, if the strength of rock is at most 1 k bar, comparable to the lithostatic pressure at about 3 km, we must expect the upper part, about 10 km, of a lithothermal system to manifest splitting and fracturing. In other words, once the magma reaches the surface zone it will find the weakest paths and penetrate them. This disturbed zone will act as a reservoir or cistern.

TABLE 14.1

Comparison of properties of the surface zones

	Hydrothermal	Lithothermal
(1) Material	Water	Liquid rock-substance, crystals, water, ambient rock
(2) Viscosity	10^{-2} P	10^3–10^{10} P
(3) Viscosity range	Moderate	Large, depends on temperature *and* composition
(4) Discharge	Water, water and steam	(*a*) Magma (*b*) Mush: magma with crystals (*c*) Foam: gassy (water) magma (*d*) Liquid–crystal–foam (*e*) Fluidized solids Water always as a gas
(5) Scale, length	100 m	10 km
(6) Permeability	Low below a few kilometres	Self-made, available at all depths
(7) Flashing	Boiling-point met at 10^2 m	Degassing possible at 10 km, occasionally down to 100 km.
(8) Scale, time	Rapid: hours, days	Years

The nature of the working fluid

Whereas in a hydrothermal system the fluid is solely water, in a lithothermal system we find a wide variety of magmas. But by far the most voluminous of these are of two kinds: (1) granitic magmas which are wet, have high viscosity and a very narrow solidus–liquidus temperature range; (2) basaltic

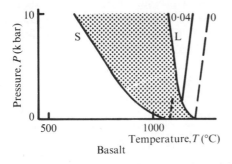

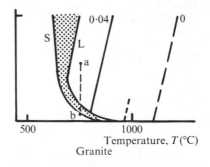

FIG. 14.2. The state of the rock-substances basalt and granite for various pressures, *P* (kbar); and temperatures, *T*(°C). The shaded region shows the range of conditions for partial melting of wet material. Below the curve S, the so-called saturated solidus, all the substance is solid; above the curve L, the so-called saturated liquidus, all the substance is liquid. These curves are for a wet system. The corresponding lines for a dry system are shown as dashed lines, the liquidus labelled 0. An intermediate liquidus for 4 per cent water content is shown. (After Harris.)

magmas, which are nearly dry, have low viscosity and a broad solidus–liquidus temperature range. The gross properties of the respective volcanic systems are determined by the nature of the working fluid and, as we shall see, principally by the water content and viscosity of the magma when it enters the volcanic system.

Although the arguments are generally applicable, in order to be explicit most of the discussion of this chapter, unless stated to the contrary, assumes a basaltic system, which is much the commonest.

TABLE 14.2

Properties of basaltic and granitic magmas

	Basalt	Granite
Density	$2.65\,\mathrm{g\,cm^{-3}}$ at $1200\,°C$	$2.4\,\mathrm{g\,cm^{-3}}$ at $800\,°C$
Water content: range	1 per cent: 0·3–3 per cent	10 per cent: 2–20 per cent
Viscosity	$10^3\,\mathrm{P}$	$10^6\,\mathrm{P}$
Latent heat of crystallization	$400\,\mathrm{kJ\,kg^{-1}}$	$270\,\mathrm{kJ\,kg^{-1}}$
Melting gradient, dry	$6\,\mathrm{K\,kbar^{-1}}$	$12\,\mathrm{K\,kbar^{-1}}$
Solidus–liquidus temperature range	Large: $\sim 100\,\mathrm{K}$	Small: $\sim 10\,\mathrm{K}$
Origin	Mantle	Granitic crust
Eruptive mode in general	Quiet	Explosive

Chemical buffering of available water

When the access of free water is restricted and the magma contains constituents which could combine to form compounds involving chemically bound water, there is a potential competition for the water if portion of the magma cools sufficiently for some crystals to form. The archetypal minerals with bound water are the micas: muscovite, biotite, phlogopite, and the amphiboles, notably hornblende. The main effect of this competition is for there to be a region in which the relation of pressure P to temperature of crystallization T is opposite to normal so that $dP/dT < 0$. For example in a biotite-rich granite about 1 kbar at $700\,°C$ the water will be buffered by the biotite so that the onset of melting increases about $10\,\mathrm{K\,kbar^{-1}}$.

A dry melt, if hot enough, could rise to the surface and remain a liquid. So could a melt with excess free water as a kind of crystal-filled soup. In this second case there is the possibility that the upward rise might be terminated below the surface because the entire mass had solidified. This situation is indicated on the $P(T)$ diagram by a typical such path ab. Thus, for example, such a saturated granitic liquid will freeze at about $700\,°C$ at 2 kbar, that is, about 7 km down.

Aspects of this kind are not considered further in this book.

Evolution of a basaltic volcanic system

1. At a depth $H \sim 100\,\mathrm{km}$ partial melting is produced in the region of a positive temperature fluctuation in the upper mantle. In order for an

adequate volume of the upper mantle to be so affected the temperature fluctuation must persist for a sufficient time.

2. Either because of the uplift produced by the lower densities in the zone of partial melting and its immediate ambient or to concurrent regional tectonic movements channels will be produced to the surface. If this does not happen the zone of partial melting will return to its original state after the temperature fluctuation has ended.

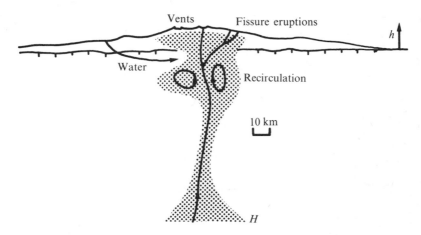

FIG. 14.3. Diagram of a volcanic system. Vertical section of an old crust punctured by a zone of spasmodic partial melting, shown dotted, together with the new pile of volcanic debris.

3. Assume that the working fluid of the system is liquid basalt, derived directly or by differentiation from the ambient material. Noting that on melting there is a fall of density $\Delta\rho$, the potential height to which the fluid column can rise above the original surface is $h = H(\Delta\rho/\rho)$. Typically $\Delta\rho/\rho \sim 0.1$, so that in this example $h = 10\,\text{km}$. Later we shall look more carefully into the problem of the potential rise in a stratified crust.

4. The fluid will not, of course, suddenly spew out in a fountain 10 km high. Work must be done against the hydrodynamic resistance to the flow in rising up the channel. Nevertheless, once the channel has been established a rapid sequence of events analogous to the opening of a hydrothermal bore or the initial phase of a geyser eruption will occur. At first the upwelling liquid will come rather slowly out of the channel but the rate will increase rapidly. The excess pressure available at this stage of fairly gentle outflow will easily be sufficient to rupture the upper crust in many places, rather like a blow on a slab of concrete. Dyke eruptions will progressively increase.

5. The initial phase will now reach the mature stage of vigorous eruption. This is the phase of so-called plateau volcanism—voluminous hot flows spreading over progressively larger areas as the original topography is filled in—analogous to the steady discharge of a hydrothermal bore. This phase will continue with progressive fall-off until the basaltic deposit is of thickness h. An interval of typically 0·1–10 million years has elapsed.

6. Two possibilities now remain.
    (a) If the permeability of the system is large enough and there is no other change owing to increased mantle or regional tectonic activity the system is dead.
    (b) If the permeability is small enough a geyser-like action is possible. For a persistent deep zone of partial melting such a case must inevitably arise through the closing of channels as the upper system cools.

7. The zone of partial melting freezes up and all volcanism in the system is over.

Spatial features: rock pile

The net effect of a sequence of eruptions is merely to make a pile of rock-substance. That pile represents the integrated effect of the volcanic system over a time that is long compared to that of an individual eruptive episode. Viewed thus, a volcanic system is close to hydrostatic equilibrium and the broadest features of the structure can be readily described.

Theoretical sketch: elevation of the lava pile
Oceanic crust. Of the many possibilities, consider the case of an oceanic crust of thickness c and density $3·0\,\mathrm{g\,cm^{-3}}$ under an ocean of depth d with the base of the lithothermal system a distance H below the crust in mantle of density $3·3\,\mathrm{g\,cm^{-3}}$. Assume that the mantle fluid rises a distance ξc into the

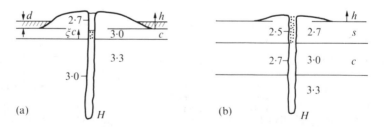

FIG. 14.4. Diagram for a hydrostatic model of a volcanic system in a stratified crust, vertical section: (a) oceanic, basaltic; (b) continental margin, andesitic. Layer and fluid densities, in $\mathrm{g\,cm^{-3}}$. Not to scale.

crust, where $0 < \xi < 1$, before it differentiates into a basaltic liquid of density $2.7\,\mathrm{g\,cm^{-3}}$ and a denser residue which remains in the lower crust or is returned to the upper mantle. Equating the pressure in the ambient rock, the so-called lithostatic load, and the lithothermal system and ignoring, for the moment, the pressure arising from the new volcanic pile (since during the early stages it will be of small extent and contribute negligibly to the pressure at depth):

$$h = \tfrac{1}{9}H\{1 + (1-\xi)c/H\} + 0.4d.$$

The contribution of the term in $c/H \approx 0.05$ is fairly small so that the effect of ξ, the level of basalt generation, on the gross features of the system is minor. For the Hawaiian islands with $d = 5\,\mathrm{km}$ and $h = 9\,\mathrm{km}$ we have $H = 63\,\mathrm{km}$. This is comparable with the depth of the deepest volcanic microseisms.

Undoubtedly $\xi > 0$. For example, studies of the gravity field in the Hawaiian islands reveal localized gravity 'highs' of about 100 mgal. This would require up to 8 km thickness of mantle material within the crust and the volcanic pile. In our model this material arises from two sources: the dense residual magma left behind after the generation of the basalt, the cooled portion of extent ξc of frozen mantle material. Seismic studies have found material of velocity $7.6\,\mathrm{km\,s^{-1}}$, comparable to mantle material.

An analysis of this kind assumes that during differentiation the bulk of the heavy residue, which is to some degree a crystalline mush, remains trapped in the spaces created in the ambient rock by the invading material, and does not thereby contribute to the pressure in the system.

Continental crust. If we apply the oceanic model to andesitic volcanism within a continental crust we obtain a similar set of relationships. But these volcanic piles typically reach thicknesses of about 3 km rather than 10 km. Thus we need to search for a model of about a third the vertical extent. Alternatively, if we consider a layered continental crust of thickness $s \sim 15\,\mathrm{km}$ of $2.7\,\mathrm{g\,cm^{-3}}$ and lower crust of thickness $c \sim 15\,\mathrm{km}$ of $3.0\,\mathrm{g\,cm^{-3}}$ and a model like that for the oceanic crust, the best fit is for $H \sim 0$. Thus, taking the extreme case of zero penetration of the mantle lithothermal system into the crust and the densities of the melts of lower and upper crust to be $2.7\,\mathrm{g\,cm^{-3}}$ and $2.5\,\mathrm{g\,cm^{-3}}$, $h = 0.04(25 + 3c)$. For the values above, then $h \approx 3\,\mathrm{km}$.

We deduce that andesitic volcanism relies on the direct use of continental crust and does not arise by direct flow and differentiation from the mantle.

Theoretical sketch: the shape of the basaltic lava pile; an estimate of strength For the opening or closing of channels we have used a simple criterion of a finite strength. Let us make the assumption that the shape of the lava pile adjusts itself so that the finite strength is reached equally throughout the pile. If the pile is in equilibrium, then from the balance between the stress on the base and the pressure gradient arising from the weights of adjacent columns of different height we find

$$h/r = 2\tau/\rho gh,$$

where h is the central height above the base, r is the radius of the base, and τ is the finite strength. For example, for Hawaii with $h = 8\cdot7\,\mathrm{km}$ and $r = 85\,\mathrm{km}$ we deduce that $\tau = 0\cdot13\,\mathrm{kbar}$, which is very reasonable.

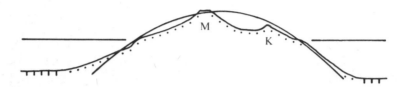

Fig. 14.5. A lava pile: a vertical section through Hawaii, drawn between Mauna Loa, M, and Kilauea, K. Vertical exaggeration 5:1. On the flanks, the ocean and oceanic crust. For comparison: the section through a paraboloidal cap of height $8\cdot7\,\mathrm{km}$ and width at the base of $170\,\mathrm{km}$ corresponding to the size and shape of a pile of finite strength $0\cdot13\,\mathrm{kbar}$.

Thus we have a method of finding the total mass that could at most be erupted from a localized region if erosion were negligible. Statical considerations give the potential height of the pile. The requirement of finite strength gives us its potential shape. Hence we obtain the mass,

$$M = \rho^3 g^2 h^5/6\tau^2,$$

where
$$h = H(\Delta\rho/\rho).$$

Notice how strongly M depends on H: the bulk of the mass output comes from the deeper reservoirs.

In the early vigorous stage erosion cannot keep pace with the rate of production of new surface material. But as the system gets old and tired erosion can allow a brief final spurt. As erosion lowers the surface elevation of the volcanic pile the diminished head of fluid may again be sufficient to extrude material on to the surface. This residual magmatism is especially noteworthy in the late-phase eruptions of Hawaiian type.

Theoretical sketch: the age of a lava pile

Finally, let us estimate the time required to make a lava pile from the following model.

1. A mantle pump at pressure P, determined by the lithostatic load, feeds the pile through a fixed resistance R. Thus the rate of mass injected into the pile $m = \rho(P-p)/R$. If the mantle pump is at depth H and the pile height is h, then $(P-p) = h_\infty - h$, where $h_\infty = H\,\Delta\rho/\rho$ is the greatest possible height of the pile.

2. Throughout its growth the pile exhibits finite strength, as described above. Thus the mass of the pile M, a known function of h, changes at the rate $dM/dt = m$ for negligible erosion.

This little equation is readily integrated to give the time t to reach the height y in the form $t = t_0 S(\xi)$, where t_0 is a time-scale and $S(\xi)$ is a dimensionless function of the relative height $\xi = y/h_\infty$ such that

$$t_0 \approx 5\rho^2 g^2 R h_\infty^4 / 6\tau^2,$$
$$S(\xi) = \tfrac{1}{5}\xi^5 + \tfrac{1}{6}\xi^6 + \ldots.$$

Inserting a typical set of values: $R = 10^3$ (km-head) d km^{-3}; $h_\infty = 10$ km; $\tau = 0.13$ kbar, we find that the time-scale t_0 is about a million years. Whence we obtain

time (10^6 yr) :	0·01	0·1	1	10	30
height (km) :	3·3	4·9	7·2	9·4	9·9

The rate of growth slows down very quickly. Two effects contribute to this: the diminishing available pressure difference $(h_\infty - h)$ and the proportionally greater mass required to expand the pile as it gets bigger. The first few kilometres can grow in a time of the order of 10 000 years. Our assumption of finite strength will be rather poor at this stage; the pile will be more like a jumble of individual calderas and their associated lava flows, many of which will be voluminous enough, owing to the large head available, to flood extensive areas. After a few million years the system has nearly choked itself off. If the condition of the mantle pump does not change, the pile will, however, continue to be active for a time of the order of 10 million years. The ultimate fate of such a pile, if its age is comparable to the erosion time-scale, would be an equilibrium between the mass input to the pile and the mass output due to erosion.

Estimate of melt proportion

The basaltic rocks of the Karroo, which were erupted about 150 million years ago, inundated an area of about 2×10^6 km^2 with lavas and swarms of dykes and sills of rather uniform composition, corresponding to conditions at about 10 km in depth. This event occurred during the prelude to the vigorous rifting and during a general upwelling of the crust of at least 1–2 km. The effect of erosion is included in this figure, so let us take 1 km of crust as the working head of the system. We can then estimate the gross proportion of melt required, assuming that the depth of the system was 100 km, and the density ratio melt/solid was 0·1, to be about 10 per cent. This is a rather porous system!

Sills

The elevation in the crust of the top of a lithothermal system may not be sufficient for a surface discharge. Nevertheless, with typical sediments of $2.4\,\mathrm{g\,cm^{-3}}$ and basement material of $2.7\,\mathrm{g\,cm^{-3}}$ there is the possibility that some or all of the sedimentary cover may float on the invading lithothermal

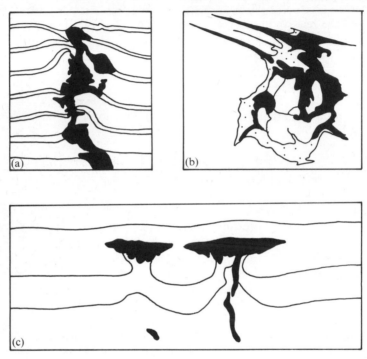

FIG. 14.6. Laboratory models of intrusive processes: (a) magma penetrating and distorting a layered crust; (b) a collapse structure; (c) sills. (Tracings from photographs, after Ramberg.)

system. Undoubtedly this is the origin of some of the enormous dolerite sills, some of which cover an area of the order of $10^4\,\mathrm{km^2}$. Clearly the thickness of the sediments to be floated needs to be sufficiently great for the difference between the pressure in the lithothermal system and the ambient rock to exceed the breaking stress of the beds. This could be less than $0.1\,\mathrm{kbar}$, corresponding to a depth of floated layer as little as a few hundred metres. Under these circumstances some fracturing of the upper sediments may occur but these fractures will not be invaded by magma.

In the early stages of the development of the volcanic system in a sedimentary area the formation of sills will thus be common. But extensive floating of light crust will no longer be possible as the lava pile builds up: the magma is less dense than its own solid.

Temporal features: the volcanic clock

The consideration of the hydrostatics of a volcanic system allows us to describe and identify the gross features such as the depth of the system and the height of the ultimate lava pile. But that is only part of the story since it ignores the essentially dynamical nature of a volcanic system.

The most striking feature of a volcanic system is not that matter is discharged at the earth's surface; it is that the discharge is pulsatory. In old volcanic areas we can find a multitude of layers, each clearly identifiable as a distinct lava flow, often one above the other. Such 'layering' is found both in basaltic areas and in areas of granitic volcanism. This pulsatory behaviour is, of course, well known from both historical and present-day observation. Brief intervals of vigorous eruption are followed by longer intervals of no apparent surface activity. These pulses are not evenly spaced in time but only roughly so for a particular system.

What and where is this somewhat erratic clock? How could we set about building a clock on the 100 km scale that would tick, say once every 10 years or so, like the clock under Hawaii? As we shall see, there are several clock-like elements possible in a volcanic system. It is, however, worth noting that we are going to take ideas from studies of geysers and simply apply them to volcanoes. We shall begin with the clocks which dominate basaltic volcanism.

Reservoir model of a volcanic clock

A clock is a resettable interval-timer. A simple timer could be constructed from a can with a narrow tube set into the bottom, which is filled to the top at the beginning of the interval. The end of the interval could be taken as the instant the can is empty. If then the can is refilled a sequence of equal time-intervals is defined. If the total outflow is collected in a bucket the total time is represented by the amount in the bucket. The time-interval can clearly be changed in two ways. (1) Changing the tube will allow control over the rate of discharge. (2) Changing the area of free surface will alter the rate at which a given discharge empties the can. We can specify quantitatively these two elements of the interval-timer as the capacitance of the can and the resistance of the tube. If we measure pressure as a head of the working fluid, then: the capacitance C is the change in volume of fluid in the can per unit change in pressure; simply the area A of the free surface; and the resistance R is the pressure drop to drive unit volume per unit time through the hole. If the can also contains a permeable medium of porosity e, then $C = eA$. We then find that the length of the interval is proportional to the time-scale RC.

If a number of cans or reservoirs are interconnected by tubes we can make systems which have several time-scales.

Thus in the upper mantle and crust we consider a set of reservoirs, the output of one being the input of the next. These reservoirs are not great holes in the ground. They are volumes of permeable ground occupied by partially molten rock-substance.

The reservoirs and their connecting channels need not pulsate; a steady state is quite possible. In general, however, as opposed to the analogous hydrothermal systems, because of the nature of the venting system the discharge is not steady. This is discussed in detail below. For the moment we

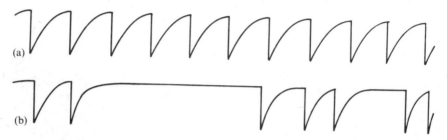

FIG. 14.7. The 'tick' of a pair of 'reservoir clocks' each constructed from three reservoirs in series, the last representing the reservoir which feeds the volcanic vent. The graph shows the level of fluid in the vent reservoir as a function of time. Time-scales RC in the ratio $10:10:1$, with nominal ratio of time between eruptions to duration of eruption of $10:1$. (a) a regular system; (b) a random system, exactly the same, except that the resistance of the feeder to the vent reservoir is reset after each eruption to a value proportional to $(1+\varepsilon)$, where ε is a random function of range ± 0.5.

shall simply assume that once a critical state has been reached in the venting system it is triggered and the vent fluid is quickly discharged. Our reservoir interval-timers can then manifest themselves and be identified.

So far we have produced a regularly ticking clock. The natural system is, however, somewhat erratic, and our model shows us that this is not surprising. We have assumed that the venting and the condition of opening and closing feed channels is fixed.

We can simulate variability in our model by choosing after each eruption a slightly different new condition for opening channels or venting. Since the system is close to choking itself off, a small variation in these conditions leads to a pronounced variability in the volcanic activity, notably in the interval between eruptions.

Mechanical trigger: pumping up the reservoir

Immediately after the eruptive phase, the pressure in the vent and adjacent reservoir is very low. Hence, as well as an over-all compaction of the upper part of the system, there will be an increase in the flow-resistance. As fluid from depth begins to recharge the reservoir there will be a gradual swelling of the reservoir and an associated decrease in the flow-resistance, continuing until the start of the eruption. This is just like pumping up a big bag, except that the opposing forces come not from the stretching of the skin of the bag but simply from its weight. The role of this heavy bag is important in all eruptive volcanism.

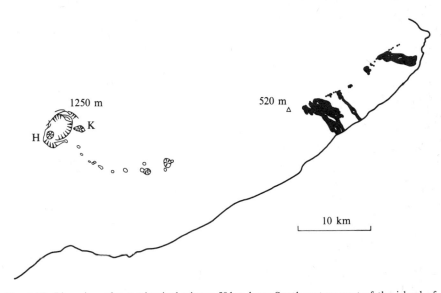

FIG. 14.8. Plan view of a mechanical trigger 50 km long. South-eastern part of the island of Hawaii showing: the Kilauea caldera at 1250 m altitude, inside, the crater Halemaumau, H, and just to the east the crater Kilauea Iki, K, and the near-by chain of craters; the local summit of Heiheiahulu at 520 m altitude; the lava-flows of the 1955 eruption; the line of spatter and cinder cones at vents of the 1955 eruption.

The sequence of events was:

1923–54	Quiet.
30 March 1954	Two earthquakes, 20 km under rift. Subsequent earthquakes under rift and caldera. Mountain swelling.
31 May 1954	Brief eruption in lava lake, Halemaumau. 0400 h, Lake filled to depth of 20 m in 2 hours. 1200 h, 50 per cent of lava drains from lake back into the vent. Activity ceased after 3·5 days.
February 1955	Greatly increased earthquake activity.
28 February 1955	0800 h, start of flank eruption.
7 March 1955	Varying activity in several vents, each lasting a few days. Subsidence in caldera starts and continues for about a year.
27 May 1955	Eruption finished. Total discharge 0·1 km^3.
14 November 1955	Next eruption starts, in Kilauea Iki crater.

If the pressure in the bag exceeds the lithostatic load by an amount equal to the finite strength of the rock, fracturing can occur. Part of the bag fluid will then be squeezed out until the pressure in the bag falls below the lithostatic load, whereupon the discharge ceases and the bag begins to fill up again.

The volcanic piles produced by basaltic and granitic lavas differ in one notable respect. The less stiff relatively weak basaltic pile is easily broken so that flank eruptions are a typical feature. The granitic pile, by contrast, is relatively stiff and strong; flank eruptions are less common, and so the system has to keep working away until the central vent is filled. The action of filling the central vent is like the man who is paying off a mortgage that is barely within his income. Only a very slight increase in the mortgage rate can greatly extend the period needed for repayment. The opportunity to have a flank eruption is like winning the pools.

I have simulated a system like that at Kilauea, Hawaii, where flank eruptions are usual. This is an excellent example of a system with a mechanical trigger.

Theoretical sketch: model of a dyke eruption

Let us try to construct a simple model of a system dominated by dyke eruptions and roughly calibrate it with data from Hawaii, notably the 1955 Kilauea flank eruption. Assemble the following components: a magma

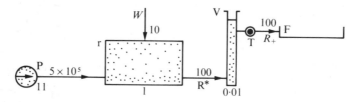

FIG. 14.9. Diagram for a model of a flank eruption. P, mantle pump; r, vent reservoir; W, weight of rock above vent reservoir; V, caldera vent system; T, mechanical trigger; F, discharge from system. Pressure and load in km-head of magma; capacitance = area of reservoir × porosity (km²); resistance in (km-head) d km⁻³. Note for the response shown in Fig. 14.10: (*i*) the reservoir internal resistance, R^* is proportional to the level in the reservoir, and is set at $R^* = 100$ when the reservoir is as full as possible; (*ii*) the discharge resistance R_+ is randomized uniformly in the range 30–300 with a mean value of 100.

source; a closed-vent reservoir; a vent; a discharge point; connected with resistances. The numerical values of the various items are chosen to fit:

(1) the elevation of the crater lake;
(2) the interval between eruptions, 3×10^3 days;
(3) the duration of the eruption, 30 days;
(4) the total mass discharge, 2×10^8 ton;
(5) ground swell and collapse of range 0·01 km.

The choice of the feed resistance, here given as 5×10^5 (km-head) d km⁻³, is of interest. For values much less than this the swelling of the ground occurs very late in the filling period. The recharge is small and the vent is filling from the vent reservoir. Only when the vent is nearly full, so that the rate of withdrawal from the vent reservoir is small, can the recharge catch up.

The trigger opens when the trigger pressure exceeds the lithostatic load by the finite strength, and closes when the pressure drops below the lithostatic load.

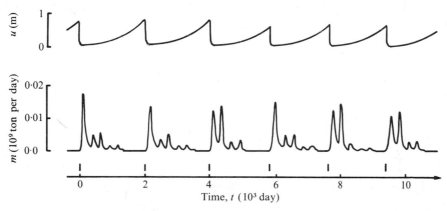

Fig. 14.10. Flank eruption model: mass discharge rate m, in 10^9 ton per day as a function of time t, in 1000-day units. The curve $m(t)$ is shown with the start of each eruption marked by a short vertical line; thereafter the time-scale is expanded 40:1 until the end of that eruption. Here each eruption takes about 30 days and gives a total volume discharge of about $0.25\,\text{km}^3$. Also shown, without time distortion, is the relative vertical displacement of the vent region u, in metres, computed as 0.01 of the vent reservoir level change.

The ground swell arises from the extra volume of the reservoir. As the reservoir fills a proportion of the ground, measured by e, the porosity is invaded by fluid—either directly or by remelting. Thus, for a level change ξ in the reservoir the ground swell $u = \xi e(\Delta\rho/\rho)$.

Phase-change trigger: magma geyser

This basaltic volcanism is rather tame, but granitic volcanism is explosively violent. What is the cause of this? One thing only: water. The degassing of a wet magma is one of the most violent events found on the surface of our planet; instantaneous power levels for a single eruptive vent can reach $10^9\,\text{MW}$. And the simple fact that a gas has a very much lower density than its liquid is the origin of the enormous lifting power of wet magma.

Consider a system made of a vent and perhaps a reservoir. Let us follow the sequence of events after the moment the discharge ceases

1. At the cessation of the discharge the vent may be nearly completely empty. Magma percolates into the void—and the void begins to fill. If there is a large reservoir the filling time may be considerable; nevertheless, the level continues to rise.

2. The water content in the magma in the vent and adjacent ground at the cessation of the discharge will be low. Initially the refilling is with degassed magma, which has been degassed and cooled during a previous discharge so that a new discharge is not immediately possible. However, as refilling proceeds, water-saturated magma from depth enters the void. With some volcanoes the ambient ground pressure is sufficient to raise the level to the

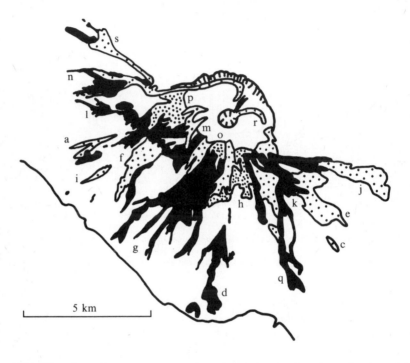

FIG. 14.11. Plan view of a magma geyser: Vesuvius, Italy, showing the major lava fields produced from 1631–1944, labelled in order (a)–(s). The present crater, the high north face of the outer crater wall, Monte Somma, and the nearby coastline are shown. (Adapted from F. Bullard.)

top of the vent and produce an overflow; in this case the saturated magma level will rise even further in the vent as more deep magma enters the void. If no overflow is possible, free convection in the vent—a slow process— gradually elevates the saturated magma level. At some level z^* the magma reaches the degassing pressure (DGP).

3. Degassing commences and fluid begins to well out of the vent. Pressures fall and degassing rapidly spreads to levels above and below the initial degassing level. The discharge for a short time rises rapidly to its maximum possible value.

The degassing pressure will be a function of the magma temperature and especially of the water content of the magma. The solubility of water-substance in a magma increases approximately as the square-root of the pressure. Clearly for a dry or nearly dry melt for which the DGP ≈ 0 the

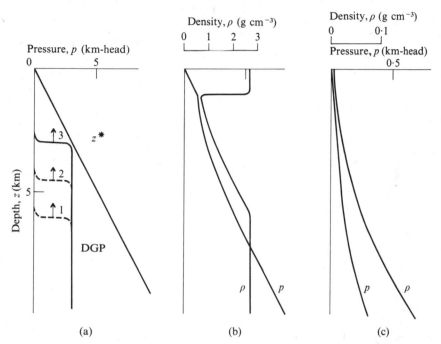

FIG. 14.12. Eruption of a wet magma from a vent. (a) Approach to eruption. The vent in the case shown here is already filled. The upper fluid is stale degassed magma. As more fresh fluid enters from the reservoir the level of fresh fluid rises in the vent by displacing stale fluid. The DGP curves are shown at successive times 1, 2, and 3. The eruption is about to commence at time 3. (b) Establishment of degassing. The density of the vent mixture and the pressure down the vent. A little stale magma at the top of the column has yet to be flushed out. The degassing zone covers already about half of the vent, and extends from the bottom of the stale magma to the DGP. Note that the pressure is reduced throughout the vent below the stale magma. Time 4. (c) Degassing fully established throughout the vent. Pressure and density down the vent. Note that the pressure and density scales have been magnified, both p and ρ are very small. Time 5. p, pressure, in km-head; ρ, density, in g cm^{-3}; z, depth, in km. Curves obtained for: 10 per cent water content; magma density 2·7 g cm^{-3}; magma temperature 1100 °C; vent area of cross-section 0·001 km^2. (Analogous to the Icelandic volcano, Hekla.)

type of geyser-like system discussed here is not possible. The only possibility is a spring-like system.

At any pressure less than the DGP the solubility of water-substance in the magma is less than that in the deep, fresh magma. A phase separation must then occur into the two phases: liquid rock-substance, saturated at the lower pressure, and free water-substance, which because of the temperature will be a gas. Thus the magma becomes a foam of liquid rock-substance and bubbles of gas. Now it might be thought that further segregation of the liquid and gas would occur. But no, there is not time, and in any event the severe turbulence in the discharge will act as an efficient mixer. It is really important to realize that the bubbles do not move separately out of the magma. Thus we can estimate the density of the liquid–gas foam by adding up the contributions of liquid and gas. As a piece of magma moves up the vent the pressure on it continues to fall, more gas is available for bubbles, and the density falls further. A few per cent of gas is sufficient to lower the density greatly.

Why does a maximum occur in the output? Whereas the thrust builds up as degassing increases in a fluid which is still fairly dense, once the degassing has reached an advanced stage the fluid density is now very small. The peak output normally occurs well before the entire column is degassing.

4. As soon as degassing commences, but especially after the peak has passed, the low vent and ambient ground pressure will lead to reservoir collapse, gradually cutting off the supply from depth so that the discharge must be maintained from the vent and associated reservoir.

5. If a reservoir exists at the base of the vent, the degassing level may enter the reservoir. The discharge, then maintained from the reservoir, would fall slowly. If there is no reservoir, z^* will quickly increase to its maximum depth. If the vent length exceeds z^*, undischarged, saturated magma will remain in the vent.

6. Above z^* degassing is occurring in the ground. If the permeability above z^* is very small, but is sufficiently large near or below z^*, no upper colder degassed magma can fall into the void and a weak vapour discharge is possible. Otherwise, colder degassed magma will begin to enter the vent and the discharge will cease.

Once degassing reaches the bottom of the vent or associated reservoir the main phase of the eruption is over. Whereas during the early stage of the eruption there was a relatively high proportion of rock-substance to water-substance in the discharge, at this stage the proportion of rock-substance in the discharge will be very small.

The essential point about this type of system, in contrast to an ordinary water geyser, is that the system to a considerable degree controls its own permeability. Withdrawal of the fluid from the reservoir will lead inevitably to compaction of voids as the ground settles under its own load, and in any event the inevitable cooling will lead to closing of voids by partial freezing of the melt.

Note particularly that geyser-like action of this kind occurs when the ground resistance is sufficiently *small*. For an ordinary geyser in the surface zone of a hydrothermal system this condition arises from conditions independent of the geyser itself—perhaps owing to the permeability of the original ground or because local ground movement around an existing hot spring have sufficiently lowered the permeability. But in a lithothermal system the permeability is created, maintained, and destroyed by the dynamical nature of the system itself.

With a basaltic magma of viscosity of the order of 10^3 P we see that the internal resistance of the reservoir is low enough to permit a steady operating point. Of course, as we have seen, owing to the inflation and deflation of the reservoir the internal resistance is modulated; nevertheless, persistent discharging is possible. With a granitic magma of viscosity of the order of 10^6 P the situation is quite different. The internal resistance of the reservoir is now so high that the operating point of the system corresponds to a collapsed state. Thus it should not be surprising that only granitic volcanoes are invariably magma geysers.

Theoretical sketch: volcanic vent output

In order to evaluate the operating characteristics of a central eruption let us make the following assumptions.

1. The system is a void, the vent, penetrating a permeable reservoir.

2. The pressure P, in the reservoir, is given by the lithostatic load, resulting from the weight of solid rock above it. The mass output m' of the reservoir, is determined by Darcy's law, as with our reservoir clock: $m' = \rho_m q$, where ρ_m is the density of the deep magma and q the volumetric discharge rate is given by $q = (P - p)/R$, where p is the pressure at the outlet of the reservoir and R its internal resistance.

The reservoir internal resistance is a phenomenological parameter of these models. We can make an estimate for any given situation provided we have some idea of the over-all internal geometry. For example, for a basaltic volcano like Hekla, Iceland, we find that $R = 100$ (km-head) d km^{-3} corresponds to a vent system of diameter 0·1 km reaching 10 km into a reservoir of permeability 30 darcy. This permeability, for a porosity of 0·1 implies that the reservoir is perfused with channels of width 1 mm.

3. The pressure in the vent will be determined by the flow in the vent, using the relations shown in the table. The main assumption is that the buoyancy of the degassing fluid is balanced by the hydrodynamic drag

4. The over-all constraints on the system are the conservation of mass of rock-substance and of mass of water-substance, and the requirement that the pressures evaluated down the vent, and through the lithostatic load and reservoir, shall be the same.

For illustration I have chosen a typical set of parameters. In a complex system of this kind with many parameters there are innumerable possibilities. Only the very simplest are referred to here.

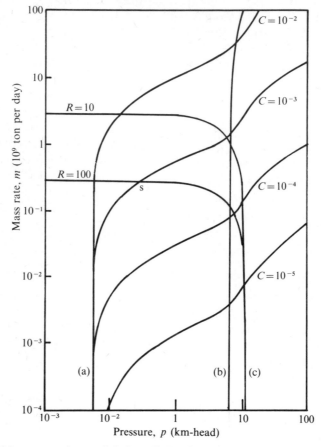

FIG. 14.13. Mass output characteristic for a discharging vent. m, mass output rate, 10^9 ton per day; p, pressure at base of vent, km-head of unsaturated magma. (a) Vent characteristic for wet magma of 10 per cent mass of water-substance for various vent cross-sectional areas, C, in km^2; from the top of the diagram: 10^{-2}, 10^{-3}, 10^{-4}, and 10^{-5}. (b) Vent characteristic for dry magma of 1 per cent mass of water-substance. Only the line of minimum working pressure is shown; to the left of this line there is no discharge. (c) Reservoir characteristic for a system with total available head of 11 km for reservoir resistances, R, 10 and 100 (km-head) d km^{-3}. Deep fluid density $2 \cdot 7\,g\,cm^{-3}$. The point (s) would be the steady operating point for $R = 100$, $C = 10^{-3}$.

At the outset the really striking thing is the dominant role of water content. If we compare curves (a), (b) (Fig. 14.13) we see at 1 per cent water content little opportunity for violent central eruptions, whereas at 10 per cent water content there is the possibility of staggeringly violent eruptions: the equivalent of a 100-megaton nuclear bomb every few minutes. Some very large ignimbrite eruptions may cover the ground to depth of 10 m over 100–1000 km^2, giving a total discharge per cycle of the order of 1–10 km^3.

Spectacles on that scale are at the top of a very wide range of possible outputs. A more typical rather small eruption may discharge at rates of the order of 10^{-4} km^3 per day for an interval of a few days, affecting a reservoir volume of porosity 0·1 of the order of 10^{-2} km^3. It is of interest to compare these figures with those of an ordinary water geyser which may discharge a mere 10 ton per cycle.

In the discussion so far I have assumed that the degassing rock-substance remains liquid. If not (and that is quite possible), and the rock-substance solidifies in the reservoir, the eruption is off. But if this occurs in the vent it will make little difference. The dynamics of the vent during the eruption is utterly dominated by the gas and it does not matter if the rock-substance is liquid, thereby producing a foam, or solid, thereby producing a fluidized ash.

TABLE 14.3
Vent relations for eruption of wet magma

Pressure in vent fluid	p
Mass discharge per unit time	m
Water mass proportion in vent mixture	$\xi = \xi_0\{1-(p/DGP)^{\frac{1}{2}}\}$
Gas density	$\rho_g = kp$
Density of vent mixture	$1/\rho = (1-\xi)/\rho_m + \xi/\rho_g$
Viscosity of vent mixture	$1/\rho\mu = (1-\xi)/\rho_m\mu_m + \xi/\rho_g\mu_g$
Fluid velocity	$w = m/\rho C$
Reynolds number	$(Re) = \rho w\,d/\mu$
Drag ratio	$\lambda = \lambda_\infty + 32/(Re)$
Pressure increase in dz	$dp = \rho\,dz(g+\lambda w^2/d)$

Standard values used here: $\xi_0 = 0\cdot1$, $k = 0\cdot18$ kg m^{-3} bar^{-1} at 1100 °C, $\lambda_\infty = 0\cdot01$.

NOTES

C, the vent cross-sectional area, in the data quoted here, is independent of depth z. λ_∞ is chosen for a very rough vent. p in km-head of deep magma. (Re) is a very minor parameter, since most systems operate at very large Reynolds number, when $\lambda \approx \lambda_\infty$. d is the so-called hydraulic diameter. For a circular vent it is the diameter; for a fissure the width. Subscript m refers to the rock-substance fluid.

15. Crustal boilers

WE HAVE SEEN the powerful effect of the presence of water-substance in wet-magma eruptions. But this is just one example of the role of water-substance in crustal systems. With few exceptions water is available throughout the upper parts of the crust and there are only a few geological systems in which

FIG. 15.1. Boiling mud pool, Pozzuoli. (Photograph by courtesy of T. Huntingdon.)

its role is negligible. There is, however, a wide class of phenomena in which the role of water-substance alone is dominant: so-called hydrothermal systems. These are crustal systems in which a body of hot water and rock is maintained by the circulation of water in the crust to depths of the order of 10 km.

Gross features

In order to be specific, I will concentrate attention on the hydrothermal systems found in the Taupo district of New Zealand. The Taupo thermal

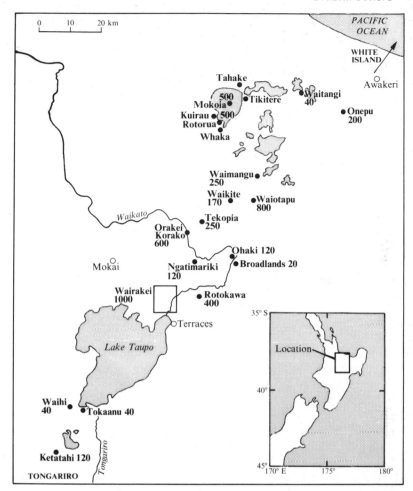

Fig. 15.2. Location and heat flows in MW of the Taupo hydrothermal systems.

areas of New Zealand occur in a zone running NNE between the Tongariro and White Island volcanoes. This hot patch lies in an area of otherwise normal heat flux. Within this zone geological, gravity, magnetic, and seismic studies reveal a depression 5 km deep filled with broken block structures, penetrated by rhyolitic volcanic complexes and volcanic debris. The total surface heat flow is 5000 MW. Included in the area of 2500 km^2 are two intense groups, Waiotapu and Wairakei, with flows of the order of 1000 MW; four moderate groups, Rotorua, Tikitere, Rotokawa, and Orakei Karako, with flows of the order of 300 MW; seven small groups, with flows of the order of 100 MW; and four very small groups, with flows of the order 30 MW. The regional average heat flux is 2 W m^{-2}, with values averaged over the more intense parts of each area of the order of 500 W m^{-2}.

The Taupo thermal area needs, for a lifetime of a million years, a supply of energy equivalent to that in a column of rock of depth 100 km and temperature excess 500 K, together with a transport mechanism to transfer the energy from the rock at depth to the base of the Taupo hydrothermal systems.

Convection of water in the crust

Scattered widely over the earth's surface there are many weak hot springs. These springs are not physically different from ordinary cold springs except that, because of the greater depth of penetration of the water, it has become heated. This heat does not necessarily come directly from volcanism but comes simply from the normal heat flux through the ground. Hot springs of this type are by far the commonest.

The situation in volcanic areas is quite different: the hot springs and other water discharges are numerous and intense. Near the surface there is a high thermal gradient, which could arise from a slowly moving vertical current of hot water which is cooled near the surface by conduction of heat to the surface. At this point the broad features of a spring-type or single-pass hydrothermal system are recognized.

There is also the possibility of free convection of water in the water-saturated permeable rock. Much of the discussion of Chapter 13 on lithothermal systems can be applied here, with two important simplifications: the working fluid is conserved in its passage through the system, and the permeability is determined by conditions in the ambient rock and not by the working fluid. This view is of a hydrothermal system based on the continuous circulation of water.

Identification of the elements of a hydrothermal system

A hydrothermal system is a heat-transfer mechanism in the earth's crust relying for its operation on the transport of water, but not necessarily the discharge of water at the earth's surface, and producing at the surface a so-called thermal area in which the heat flow is *different* from normal.

The bulk of the phenomena exhibited by a hydrothermal system can be described merely in terms of a reservoir of hot water and rock formed at depth at some previous time. The total heat flow would be considered to arise solely from the energy stored in the reservoir.

In the natural state the water-level is generally found to be stationary, and yet fluid is continually discharging from the reservoir: 5×10^3 kg s^{-1} for all the Taupo area. A water recharge system must then exist. The recharge water may be meteoric, it may originate at depth, or it may be a mixture of both. There is strong evidence on chemical grounds that the water is meteoric.

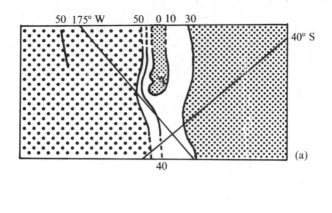

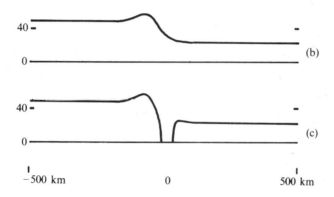

FIG. 15.3. Regional *conductive* heat flux in mW m^{-2} near the Taupo hydrothermal systems. (a) An area 1000 km × 500 km, across the thermal zone. Wairakei is shown as W. (b) Transverse profile, in the south-west and outside the hydrothermal zone. (c) Transverse profile, across the hydrothermal zone. *Note*: (*i*) the over-all pattern of two levels at 50 mW m^{-2} and 25 mW m^{-2}; (*ii*) the 200-km transition zone, lying somewhat NW of the thermal area; (*iii*) the 30–40-km strip of zero regional flux within the hydrothermal zone. It is within this region that mass discharge of hot water and vapour produces a *convective* heat flow typically 500 W m^{-2} (Adapted from Studt and Thompson, and the author.)

The temperature distribution with depth at Wairakei, obtained from several hundred boreholes, is well known to a depth of 500 m, poorly known from 500 to 1500 m, and unknown below 1500 m. But it reveals a body of water-saturated rock at 200 °C or more, 4 km wide and 1·4 km deep.

From the known temperatures, the total energy stored in the Wairakei system at the moment can be estimated to be about 10^{19} J equivalent to 20 km^3 of rock at 200 °C. This would be exhausted at the present discharge rate of 1000 MW in 250 years. There is, however, little doubt that no gross changes in temperature have occurred during historic time. Further, there is

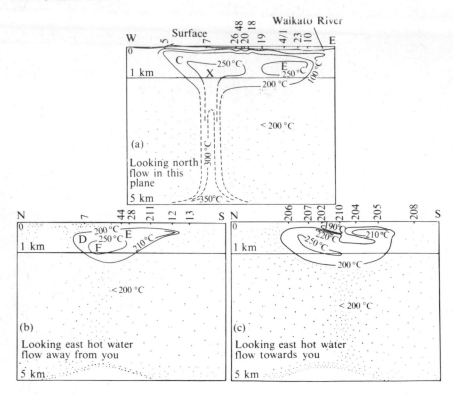

Fig. 15.4. The 'mushroom' of a hydrothermal system. Temperature distributions for various vertical sections. These are shown as lines (a), (b), and (c) in Fig. 15.8.

geological evidence of volcanic and hydrothermal activity for a million years. Thus the reservoir must be continuously supplied with energy from depth, although the rate of supply need not be constant.

Flow visualization of a simple hydrothermal system

A scale model can be constructed, using a Hele Shaw cell in the manner described in Chapter 13. The temperature distribution, measured with a small thermocouple probe, reveals strong vertical thermal gradients near the heater and the upper surface. Above the heater is a mushroom-shaped distribution. The vertical temperature profile $T(z)$ varies with position. At $x = 0$ the profile is monotonic, with strong gradients near $z = 0, 1$ but otherwise nearly constant. Elsewhere the profiles have a maximum with depth where the point penetrates the head of the mushroom. The velocity distribution reveals a plume rising above the heater. Clearly the bulk of the heat is carried in this plume. This model has zero discharge.

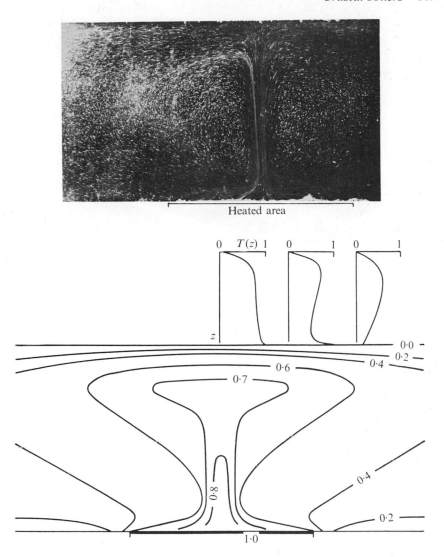

Fig. 15.5. Flow and temperature distribution in a laboratory model of a hydrothermal system. Porous medium, Rayleigh number $(Ra)_m = 10^3$.

Now let fluid be continuously withdrawn from the cell by a siphon from a small hole in the wall and at the same time let cold fluid be replaced at an equal rate at the ends of the apparatus. The temperature distribution and flow pattern depends on Q'/Q, where Q' is the power withdrawn by the siphon and Q is the total power transferred. The most striking observation is the sharp boundary between the cold recharge water and the recirculating hot

(a)

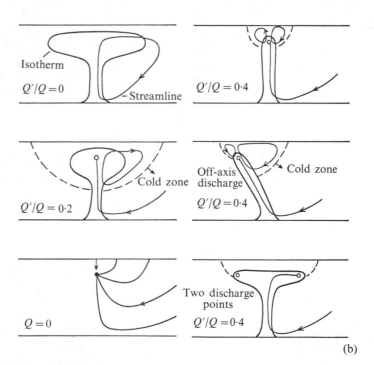

(b)

Fig. 15.6. (a) Free convection in a laboratory hydrothermal system with mass discharge and recharge. (b) Influence of discharge on the temperature and velocity distribution: (a) $Q'/Q = 0$, 0·2, 0·4; $Q = 0$; $Q'/Q = 0·4$ with the discharge point displaced, and with two discharge points. Q is the power input, Q' is the power output of the fluid discharge.

water. At small discharge rates there is little change in the temperature distribution, but as the rate is increased the mushroom collapses to a mere column of heated fluid. The situation is now largely one of a single-pass system, but we notice that the temperature distribution near the heat source is hardly altered. Indeed, throughout the experiment the power input Q remained independent of Q'. This demonstration shows strikingly that the heat transfer is determined by conditions near the heat source.

Application to the Taupo area

The field isotherms are centred below the surface thermal areas and are considerably simpler in form than the surface distribution. This pattern is just what would be expected from a section across the upper part of a mushroom which is distorted by outward flowing lobes of hot water. Provided that the system is in a steady state, only a system dominated by mass transport (here the flow of water) could produce such distributions; water rises vertically from depth until in the upper 1 km it runs away horizontally.

The most striking evidence for convection is the pronounced occurrence of cold water to depth of the order of 1 km around the immediate margin of the mushroom. It is clearly demonstrated in the model experiments that when fluid is continuously discharged the region in which recirculation of hot fluid occurs is surrounded by cold meteoric water. These observations indicate at Wairakei, for example, that $Q'/Q = 0.96$, that is, 96 per cent of the heat is discharged by mass flow, and demonstrate that Wairakei is close to being a single-pass system.

Wairakei lies within the Taupo depression, which is filled with water-saturated volcanic debris to a depth of the order of 5 km. Extrapolated borehole temperatures suggest a temperature of the order of 400 °C at the base of the depression. The heat flow by conduction alone, with a temperature gradient of 400 K per 5 km, is about 100 mW m^{-2}, very much smaller than the average value 2 W m^{-2}. Clearly, convection is the dominant mode of heat transfer.

In situ values of k for the upper 1 km of the system have been found from borehole tests to be in the range 0.1–0.5 darcy, a possible value at depth is therefore of the order of 0.1 darcy. Hence, with (a best estimate) $T = 350$ °C, and system depth of 7.5 km, the Rayleigh number $(Ra)_m = 1.0 \times 10^4$, and heat-transfer ratio $(Nu) = 250$. These are reasonable values. In particular, it has been shown that free convection in a porous medium can transport the large heat flows found in thermal areas. Incidentally, values of (Nu) could in some cases well be of the order of 10^3; bodies of kilometre scale which would cool in 10 000 years by conduction can thus cool in periods of the order of 10 years by convection.

Oceanic thermal areas

For a hydrothermal system to be possible, a source of heat and water is required, together with a layer of sufficiently high permeability. The high heat flows on ocean ridges certainly indicate a source of heat. Sea-water, rather than meteoric water, is available. Here there is the possibility of progressive closing of permeable paths by deposition of salt, but this is likely only where vaporization of water occurs. The chief problem is the question of the permeability. Permeability will be a rapidly decreasing function of

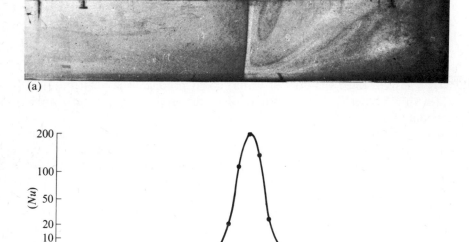

(a)

(b)

Fig. 15.7. (a) Laboratory model of an oceanic hydrothermal system. (b) Local heat flux in a model of an oceanic hydrothermal area. Porous medium, Rayleigh number $(Ra)_m = 2 \times 10^3$.

pressure, owing to the compaction of the voids in the rock. Nevertheless, in a region disturbed by surface volcanism permeabilities in the upper few kilometres of the ocean crust could be of the order of 0·01 darcy. This would be enough to allow local heat flows of the order of $500\,\text{mW}\,\text{m}^{-2}$. The permeability is somewhat lower in oceanic areas; this is already clearly indicated by the observations of lower maximum values of heat flow in oceanic thermal areas compared with land thermal areas.

The boundary conditions at the surface of an oceanic hydrothermal system differ from those on land: there is no air–water interface, and the sea-water exists both above and below the surface. This can be simulated in a laboratory model by joining the upper surface of a Hele Shaw cell into a wide cavity representing the ocean. The distribution of surface heat flux obtained in such an experiment shows that on the margins of the area the heat flux is zero, corresponding to the region of recharge of cold sea-water. In the middle of the area is a region of intense heat flux.

It is not proposed that all the heat discharged in the oceanic thermal areas is transferred in hydrothermal systems. But it should not be overlooked that some of the features of oceanic thermal areas are explicable in terms of hydrothermal systems.

Surface zone of hydrothermal systems

The surface heat-flow mechanisms, especially those found in land thermal areas, all arise in the upper 10–100 m of the crust from a body of hot water at depth, by evaporation or flashing of steam. There is an important distinction between the surface discharge mechanisms of land and oceanic thermal areas. On land, evaporation can occur at the air–water interface at the water-table; this possibility does not exist in the oceanic areas. Further, the sediment on the ocean bottom is much more homogeneous than the rock near the land surface. It is therefore to be expected that on land the surface heat flux in thermal areas will be extremely patchy, but in the ocean the heat flux will be more uniform, with much smaller local maximum values.

The behaviour of the fluid discharged at the surface in a thermal area is influenced by two factors.

1. The discharge may proceed by direct flow of the fluid of the discharge system to the surface without change of phase. These wet and dry passive spring-type discharges are dominated by the flow rather than by the presence of the surface, which is merely the level at which the discharge occurs, and their features are those of the discharge system at depth. Similarly, surface volcanism can be regarded simply as a magma spring whose properties are essentially those of the system at depth.

2. If the fluid in the surface zone is water, the discharge may be dominated by the phase change of water to vapour or steam. This may occur by flashing of water to steam within the body of a volume of water that is hotter than the surface boiling-point or by the evaporation of vapour at a water–air surface, either at the ground surface or at depth, and not necessarily at the boiling-point. Both of these processes occur independently of the level of the water-table, whereas for a spring the water-table must be at the surface. Evaporation will always occur; discharges in which flashing is dominant are called 'flashing-type'; those where it is negligible 'pool-type'.

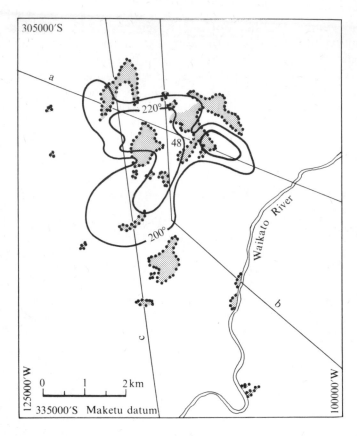

FIG. 15.8. Plan view of the surface zone of a hydrothermal system. Hot areas at Wairakei, New Zealand, with heat flux greater than 500 W m^{-2}. Superimposed are the 200 °C and 220 °C isotherms at 350 m below datum (about ground-level). The lines (a), (b), and (c), refer to the vertical section locations of Fig. 15.4. Total heat flow from this area is about 1000 MW.

The interaction of flow and phase change leads to a sequence of increasingly intense discharges:

(1) warm ground—weak steaming ground marginal to the intense areas and running out to cold ground of zero gradient and more distant normal ground;

(2) steaming ground;

(3) dry fumaroles;

(4) surface pools without overflow;

(5) springs—wet or dry (slightly superheated fumaroles) with continuous discharge;

(6) geysers—intermittent wet fumaroles and mud volcanoes;

(7) wet fumaroles—bores and mud pools with continuous discharge of wet
steam;

(8) hydrothermal explosions and eruptions.

(1)–(4) are pool-type; (6)–(8) are flashing-type discharges; but springs, though dominated by the flow, can be strongly affected by both evaporation and flashing, so that they are a combination of pool-type and flashing-type discharges. The bulk of the energy lost from a thermal area comes from pool-type discharges.

Theoretical sketch: fumarole output

Except for the evaluation of the mass proportion ξ, of steam, in the vent fluid, a fumarole can be described in the same manner as the previous discussion of volcanic vent output. Consider the following model.

1. A vent penetrates a zone of water saturated permeable rock at temperatures above the surface boiling-point. Here let us take the ground temperature $T_0 > 100\,°C$ as uniform.

2. The steam and liquid water remain in equilibrium so that the pressure equals the saturated vapour pressure (SVP). An approximation to this is an SVP of $10^{-8}\,T^4\,\text{bar K}^{-4}$, where T is the temperature in degrees Centigrade of the vent fluid. Thus as we integrate the pressure, from the surface value of 1 bar, down the vent we obtain from this relation the local temperature.

3. Field measurements confirm that the enthalpy H of the vent fluid remains constant as it moves up the vent. Thus we have approximately

$$H = H_0 + \xi\{L + c_s(T-100)\} + (1-\xi)c_w(T-100)$$
$$= H_0 + c_w(T_0 - 100),$$

where $L = 2260\,\text{kJ kg}^{-1}$ is the latent heat of vaporization, c_s is the specific heat of steam, here taken as $2\,\text{kJ kg}^{-1}\text{K}^{-1}$, and c_w is the specific heat of water. Given T, this provides a value for ξ, the steam mass proportion.

These relations for H and the SVP are quite adequate in practice, but for precise values the steam tables must be used.

For large ground resistance, the output is dominated by the ground resistance and not by the vent: the hydrodynamic forces are minor. But with small ground resistance, the output is dominated by the vent: the hydrodynamic forces are large. Thus, in the case studied here for values of $R \gtrsim 10^3$ (m-head) s m^{-3} output is nearly independent of C; for values of $R \lesssim 10$ (m-head) s m^{-3} output is nearly independent of R.

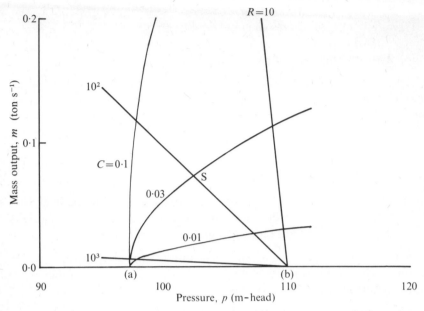

FIG. 15.9. Mass output characteristic for a discharging fumarole: m is mass output rate in tons^{-1}; p is pressure at base of vent, in m-head of hot water. Depth of vent, 100 m; deep water temperature, 150 °C. (a) Vent characteristics, with zero output at point (a), for various vent cross-section areas C, in m^2; from the top of the diagram: 0·1, 0·03, and 0·01. (b) Reservoir characteristics for various resistances R, in (m-head) s m^{-3}; from the top of the diagram: 10, 10^2, and 10^3. The point (s) would be the steady operating point for $R = 10^2$, $C = 0·03$. Natural systems, apart from a few famous ones, rarely have outputs exceeding 0·01 tons^{-1}.

Eruptive discharges

The intermittent wet fumarole known as a geyser can be regarded simply as a wet fumarole which is intermittent because of the relatively small ground permeability and whose behaviour is modified by the existence of reservoirs and ground leakage.

In mud pools, mud volcanoes, and hydrothermal eruptions, the rock takes an active part in the system. Mud is the viscous water-saturated, decomposed rock. In mud pools the steam rises slowly as bubbles rather than by the uniform permeation of ordinary steaming ground; the mud itself is almost impermeable. The flow about the bubbles will be Stokesian, and the mud transported by this flow will tend to circulate in the pool; as hot water and mud come from depth, quiet flashing produces the steam which accumulates in bubbles. Mud volcanoes are merely geysers with a superimposed mud pool; mud is thereby ejected with the flashing discharge.

Hydrothermal eruptions, so-called phreatic eruptions, are possible when water at depth approaches the boiling-point. It is possible for the saturated vapour pressure to exceed the lithostatic load there being enough thermal energy that after an adiabatic expansion to atmospheric pressure there is a large excess of kinetic energy. For example, at 250 °C water has a saturated vapour pressure of 40 bar, so that an overburden of density $2\,g\,cm^{-3}$ is in balance at 200 m depth with available energy of $50\,MJ\,m^{-3}$. If this region is extensive, an explosion is possible, ranging from pits of diameter 10 m to craters. In spite of the enormous instantaneous power developed in these explosions, the total energy released is not large compared to that of a thermal area such as Wairakei, which discharges 10^3 MW; Usu-san, Japan developed of the order of 10^6 MW and transported a total of 0·5 MW yr. These eruptions are possible in a thermal area, and it is not necessary to invoke exclusively a magmatic injection as the energy source. In a thermal area the explosions will be shallow, however, since below 500 m the temperature no longer increases rapidly with depth.

Evaporation-dominated discharges

Pool-type discharges are maintained by evaporation at a water–air boundary at the surface of a pool, in a void at depth, or in the pores and joints of the rock. It is assumed that the rate of evaporation m per unit area of interface is determined by the temperature of the water and the conditions in the air in the immediate vicinity of the interface. Experimental data for m for open surfaces in still air give $m = \sigma(P_S - P_2)$, where σ is a constant $(6\cdot55 \times 10^{-8}\,s\,m^{-1})$, P_S is the saturated vapour pressure at the interface temperature T_2, and P_2 is the partial pressure of water vapour in the adjacent atmosphere just above the interface.

Let us see how this simple idea can be applied to the New Zealand fumarole Karapiti, which (1960 measurement) discharges $4\cdot5\,kg\,s^{-1}$ of slightly super-heated steam at a temperature of 114 °C. On the assumption that the water–air interface is also at 114 °C, the saturated vapour pressure $P_S = 1\cdot63$ bar, and $P_2 = 0$, $m = 0\cdot0107\,kg\,m^{-2}\,s^{-1}$, the water area required to evaporate $4\cdot5\,kg\,s^{-1}$ at this rate is $420\,m^2$. If the steam arises within the porous rock, of porosity about 0·3, the gross area required is about $1400\,m^2$. If Karapiti were supplied with the 250 °C water of the deep part of the Wairakei discharge system and this water were not cooled or diluted by surrounding ground-water, then each gram of water at 250 °C would have enough energy to evaporate 0·25 g of steam and leave 0·75 g of water at 114 °C to leak away into the surrounding country. (Incoming 120 °C water would require an 0·99 g leak.) This leak is just part of the local convective system required to maintain the fumarole.

Steaming ground

Steaming ground can be considered to be the result of filling in a surface pool and lowering the water-table. The deeper the water-table, the less intense

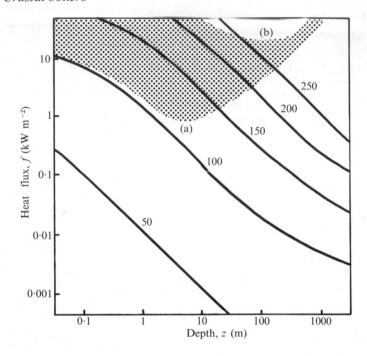

FIG. 15.10. Steaming ground output characteristic. f, heat flux ($kW\,m^{-2}$), z, depth (m) to water-table, for a variety of temperatures, 50–250 °C, at the water-table. The shaded region indicates the conditions under which phreatic eruptions may occur. Above this region steady steaming ground will be very unlikely, since the ground is explosively unstable. Curve (a): SVP = lithostatic load; curve (b): SVP = lithostatic load + finite strength (here taken at a low value of 30 bar). Data used: porosity 0·1, permeability 0·1 darcy. An output of 0·01 $kW\,m^{-2}$ would not be noticed by a casual observer; an output of 1 $kW\,m^{-2}$ is found in intense steaming ground.

the area becomes, until the upward flow of the evaporated vapour is completely impeded and the heat flow becomes conductive. Heat reaches the water-table by transfer from the hot water and rock at depth; the evaporated vapour rises through a permeable overburden and enters the atmosphere. When the vapour flow is sufficiently strong over a large area, a microclimatic effect is produced.

Assume that the vertical mass flux is em, where e is the effective porosity at the interface of the rock permeable to the vapour and P_2, the vapour pressure near the interface, is considered to be the external ambient vapour pressure P_0 plus the vapour pressure required to drive the vapour through the overburden above the interface. The role of the ambient air is completely ignored. For a shallow overburden of thickness of a few centimetres, the heat flux is given directly by em with $P_2 = 0$; otherwise it is necessary to consider the details of the flow of the vapour through the overburden by means of the equations of continuity, flow, and energy to obtain the heat flux as a function of T_2 and depth z to the water-table. For $e = 0.1$, $k = 0.1$ darcy, $P_0 = 0$, the calculations fit the Wairakei field data sufficiently well. Inspection of the maps of shallow temperature surveys, shows that the steaming areas are confined to patches within the $5 \, W \, m^{-2}$ contour; outside this contour the ground temperature will be established by heat conduction or by downward-moving cold groundwater.

A model of the Tuscan steam zone

The earliest large-scale investigation and exploitation of geothermal energy was in the Tuscan thermal area near Larderello. The Tuscan hydrothermal systems are not fundamentally different from those in New Zealand but in this model the water-table is a great depth (2 km) and the discharge

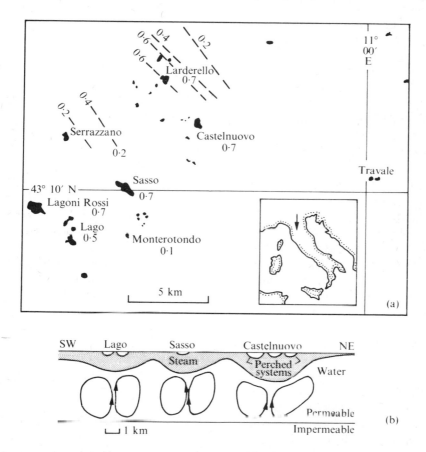

FIG. 15.11. (a) Location of the thermal areas of the Tuscan thermal district; areas indicated in km^2, surface gradient in $K \, m^{-1}$. Total heat flow about 400 MW. (b) Diagrammatic distribution of water and vapour in the Tuscan hydrothermal system. Recharge not shown.

mechanism is similar to that of steaming ground. This deduction relies principally on two sets of observations: the pressure transients, and the slight but increasing superheat during the observation period.

One of the oustanding characteristics of the Larderello project is the absence of a steady state during the last 50 years of intensive exploitation. A change in the setting of the well-head valve of a bore results in a slow change (10–100 hours) to a new steady state. Whereas the output of a single bore in an unexploited area remains constant over a period of 10 years, in an exploited area not only does the output of individual bores continuously decrease, but so does the total output; to maintain production at Larderello, a continuous drilling programme is necessary.

The half-life for an exploited area is about 10 years. Assuming the existence of a steam reservoir, an immediate explanation of these transients is possible. The mass of a compressible fluid contained in a reservoir is a function of the pressure distribution; any change in the pressure distribution will require a mass transport to adjust the mass to its new value. A transient response will occur during the time of this flow, and from the time scale of this response we deduce the thickness of the steam zone to be about 2 km.

There is another remarkable observation. Whereas the early wells at Larderello (and not just those penetrating only into the superficial near-surface groundwater zone) gave steam that was either saturated or very nearly saturated, in the last 50 years of intensive exploitation the degree of superheat has steadily risen: a well-head temperature increase of as much as 40 K has been reported. The maximum steam enthalpy at Larderello is at present about $2870 \, \text{kJ kg}^{-1}$ though the maximum possible enthalpy of saturated steam is $2800 \, \text{kJ kg}^{-1}$. At this point we must modify the view of a static steam reservoir and consider the origin of the steam. Here it is suggested that the system is a 'wet convector' but with evaporation occurring at the water surface as in steaming ground. In the original state there are good indications that the steam zone was confined by a perched water layer. In this case, $P_0 \approx P_S$, the gas enthalpy is nearly equal to the saturation value at T_2, for example, a maximum of $2800 \, \text{kJ kg}^{-1}$ at 235 °C. During exploitation it has been engineering practice at Larderello to set the bore well-head pressure at 5 bar, and, as can be shown, loss of pressure as the fluid ascends a bore is negligible, so that ground pressure P_0 has been reduced from values at least as large as 30 bar to 5 bar. Hence an order of magnitude increase in heat flow is easily possible. This implies a fall in P_2/P_S. Inspection of the steam-table shows in the region of interest, at the same temperature, an increase in vapour enthalpy with reduction in pressure. For example, the observed increase to $2870 \, \text{kJ kg}^{-1}$ for $T_2 = 235$ °C is given by a change in P_2 from 30 bar to 20 bar. It is important to emphasize that the main assumption is that the evaporated vapour in the immediate vicinity of the evaporated surface is at pressure $P_2 < P_S$ but at the same temperature as the water-surface T_2.

Pipe model

Thus the elements of a hydrothermal system are a heat source, a recharge system, a recirculation system, and a surface discharge system. Given these elements alone, it is of interest to construct a pipe model of a thermal area.

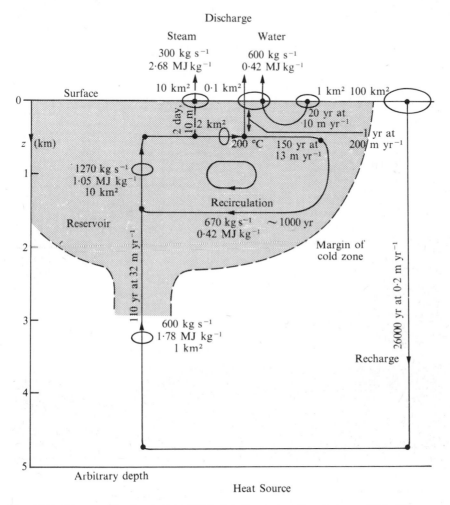

FIG. 15.12. Diagram of a pipe model of Wairakei: flow in $kg\,s^{-1}$; enthalpy in $MJ\,kg^{-1}$, cross-sectional area in km^2; velocities in $m\,yr^{-1}$, time in yr.

Our model for Wairakei is based on the 1961 data: a steam discharge of $300\,kg\,s^{-1}$ and a water discharge equivalent to $300\,kg\,s^{-1}$ of deep water and $300\,kg\,s^{-1}$ of locally heated groundwater. Included are estimated areas of the 'pipe' cross-section and the corresponding percolation velocities. Such a model is clearly extremely crude but it is valuable for calculating orders of magnitude and should be an initial step in exploring a new area. The times revealed are surprising, both for their magnitude and the large range of scales: of the order of 10^5 yr for the deep recharge, 10^3 yr for the recirculation, and 10 yr for the surface-water discharge.

Closing remark

Man's exploitation of his environment draws on three distinct resources. (1) The biosphere is in essence a chemical factory fuelled by the sun. The recycling times are geologically extremely rapid and the processes quite

FIG. 15.13. View of the geothermal project, Wairakei, Taupo district, New Zealand, mid-1961. Looking south over the main production area we see about half of the $5\,\text{tons}^{-1}$ discharge of water, in the centre distance is the Karipiti natural geothermal area; in the right distance, under cloud, the Tongariro volcanoes.

independent of direct effects deeper than, say, 100 m below the solid surface. (2) Shallow crustal deposits placed there in the remote past and in the human time-scale of finite amount and irreplaceable. (3) Interior resources contained within the at present inaccessible 99·9 per cent of the earth's mass. Only one commodity is at this time taken from this region: heat. Admittedly the heat currently drawn off, of the order of 10^3 MW, is a rather small part of human global needs, and geologically it is just another finite deposit, but in human terms it is self-renewing and permanent.

Reading list

THE FOLLOWING short list of books are chosen to provide adequate background and follow-on material. Other books and references are given in the glossary–index.

HOLMES, A. (1965). *Principles of physical geology.*
  A superb descriptive presentation. Worth buying.
CHALLINOR, J. (1967). *A dictionary of geology.* University of Wales Press.
Foundations of earth science (1968). Prentice-Hall, New York.
  Elegant coverage of a variety of broad topics.
Encyclopaedia Brittanica (1974).
  Numerous authoritative articles, the outstanding non-specialist reference work.
  See also the Bibliography of relevant sections.
CALDER, N. (1972). *The restless earth.* B.B.C. Publications, London.
  A delightful book, related to a TV documentary, which gives the flavour of an aspect of modern ideas on global geology.
Geophysical Monographs (1960—). American Geophysical Union, Baltimore.
  Specialist works of great detail but an excellent route into the raw material of current investigation.
Journal of Geophysical Research. American Geophysical Union, Baltimore.
  By far the best journal of its kind.

Questions

THESE QUESTIONS are not only to help you review the material of the book but to provoke ideas and questions of your own. Many of the questions are quantitative and should you find the detail beyond you at least try to answer them in a qualitative way by identifying the vital ingredients.

Where data are not given they will usually be found in the glossary–index. Otherwise you are expected to find them for yourself by guesswork, previous knowledge, or looking them up.

Most of the questions, in the framework in which they are posed, have definite unequivocal answers but some have no currently agreed answer.

1. You have just won a competition. Your prize is 30 minutes with the Archangel Gabriel. You can ask ten questions about the interior of the earth. Prepare your list of questions.

2. Identify approximately where in the solar system the photograph shown in Fig. 2.1 (p. 4) might have been taken, and at what time of the year. Assuming that the photograph was taken as our cosmic traveller approached the earth, suggest where it might live.

3. Bloggs, a young student at Manchester, wishes to measure the size of the earth for himself. He makes the following preparations. (1) He borrows a small sextant, accurate to better than one minute of arc, and learns to use it. (2) He borrows an expensive wrist watch which can be set and read to better than 1 s over a period of a month. (3) He fits a wheel counter to his bicycle and adjusts it by riding along a piece of straight road which he and his friends have measured using a surveyor's tape. The counter is found to give distance on the straight of ± 0.3 per cent. (4) He already owns a magnetic compass which can be read to $\pm 0.1°$.

On the 16 and 17 June he reads the elevation of the sun at noon and then sets off on his bicycle for Bath. Arriving there he repeats the elevation measurements on the 20 and 21 June and then returns to Manchester. Arriving there he again repeats the elevation measurements on the 23 and 24 June. The average difference of angle he finds is 2·04°.

On the journey, both there and back, he stops every 10 km and measures the direction of the road. He tries to be as objective as possible. On return he plots his track assuming that for 5 km either side of the stopping points he travels in a straight line. He estimates the distance between the elevation observation points to be 223 km.

Comment in detail on the accuracy of Bloggs's method. What is his estimate of the radius of the earth? How could the method be simply improved? Why not try it for yourself?

4. Bloggs heard on the radio that a new satellite especially designed for visual observation has been placed in a polar orbit at a mean height of 200 km. He decides to observe it and make his own measurement of the mass of the earth. He arranges, in his back garden, two long vertical poles spaced some distance apart, in a roughly east–west direction, so that he can have a good view of the poles in line and the sky. During a number of nights, with many blanks due to the cloudy weather, the times of passage of the satellite across the observation plane were noted to the nearest minute. The times, for each day of observation, were: 2 Feb, 1936, 2106; 4 Feb, 2111, 2243; 12 Feb, 2001, 2131, 2302; 13 Feb, 0031; 17 Feb, 2144; 19 Feb, 2017, 2318; 24 Feb, 2157. Hence, given G and that the radius of the earth is 6370 km, find its mass and mean density. Comment on the accuracy of the method, and in particular on Bloggs's time keeping. Why is the result a bit low? If the mean density is actually 5.517 g cm^{-3} estimate the actual mean height of the satellite and state the accuracy of your estimate.

5. A 10 kg steel ball is dropped down a deep vertical mine shaft. The point of impact at a depth of 1 km is found and can be located, as determined after 1000 trials, to a spot of horizontal extent ± 0.01 mm. At a level 20 m below the surface a line of loaded trucks is moved from far back to as close as possible to the shaft. The effect is a mass of 1000 ton at an 'equivalent' distance of 10 m from the line of flight of the ball.

Given the value of G:

   (a) Find the horizontal displacement of the spot.
   (b) Show how the 'equivalent' distance is evaluated.
   (c) Assess this as a method for finding G.

6. Estimate the total number of atoms in the earth.

7. Given the mass of our galaxy and the distance of the solar system from its centre estimate the time for one circuit of the solar system around the galaxy.

8. Given a knowledge of G and the distances between moon, earth, sun, and galactic centre, describe a laboratory measurement which would then allow us to estimate the number of atoms in our galaxy.

9. It is proposed to orbit a satellite around the sun within the photosphere. What will be the approximate time for one complete orbit?

10. The last Olympic high-jump champion is a bit out of condition, but though he only stands 1.60 m in his socks he can still clear the bar at 2 m. On a student field trip to the moon he is taken along. What will be his best jump there?

11. Two spherical masses m, M of material of density ρ and specific heat C in free space with their centres a distance r apart are released from rest at time $t = 0$. Under their mutual gravitational field they collide and become a single body. Estimate: (i) their velocities on impact; (ii) the time to impact; (iii) the temperature of the single body.

Take as examples the two cases, both with $m = M$:

   (a) $m = 1$ kg, $r = 1$ m.
   (b) $m = 10^{25}$ kg, $r = 10^6$ km.

12. Use the egg model and the assumptions: the earth's mass and the core-mantle density contrast have remained constant; the core radius has diminished uniformly through geological time. Deduce the earth's moment of inertia as a function of time. Comment on your result and the validity of the assumptions.

13. Given only the observed periods of free vibration of the earth quoted in Chapter 3, the mass of the earth, and the mass of Mars, predict the periods of free vibration of Mars. Estimate the lowest frequency of (a) torsional vibration (mode $_0T_2$) of Jupiter and (b) compressional vibration (mode $_0S_0$) of the sun.

14. You wish to force the entire earth into resonance so that by scanning a spectral peak you will be able to look directly for changes in the upper mantle. The method for vibrating the earth is to pump water back and forth between a large shallow reservoir and a small deep reservoir. Discuss the feasibility of this project, the type and sensitivity of the recording equipment, and how long it would be before your experiment could possibly show some results.

15. In a proposed television advertisement about chocolate we see the following. A husky individual is regularly thumping the end of a 10 km long piece of disused welded railway line, which has become loose from its sleepers, with a 10 kg sledge hammer which at the top of its stroke is 2 m above the ground. We are told that the rate of thumping has been chosen to set the rail into vibration of maximum amplitude. He keeps at it for 8 hours and then sits down and eats some chocolate. Given that the chocolate has energy content 20 MJ kg^{-1} and a single bar is 125 g, find out how many bars of chocolate he needs to consume.

Fifty years later the same commercial is filmed on the moon. How many bars of chocolate will he need then.

16. Given that g is uniform in the mantle and that the *in situ* density at the base of the mantle is 5·9 g cm^{-3}, estimate, directly from the equation of state, the density of mantle material at zero pressure. What could this material be?

17. Assuming that the effect of a uniform spherical mass is the same as an equal mass concentrated at its centre show that the vertical component of gravity measured on a horizontal surface is

$$g' = \tfrac{4}{3}\pi G \, \Delta\rho z (R/r)^3,$$

where a sphere of radius R and density $\rho + \Delta\rho$ is buried, in ground of density ρ, with its centre at depth z and r is the distance between the observation point and the centre of the sphere.

A granitic intrusion pierces unconsolidated sediments; the top of it just reaches the surface, and the maximum gravity anomaly is 4 mgal. Assume the intrusion to be roughly spherical in shape. The concentration of heavy metals in the granite is 150 p.p.m. What is the mass of the heavy metals in the entire intrusion?

18. Calculate the thickness of the continental crust in a region where the densities of continental and oceanic crust are 2·8 cm^{-3} and 3·1 g cm^{-3}, the ocean is 4 km deep, and the continental elevation is 1·4 km above sea-level.

19. An ocean basin has been stable since the Jurassic Period. The average depth of water over an area of order 2×10^5 km^2 is 2·6 km. Unconsolidated sediments cover the ocean floor to a thickness of 0·5 km. Is the crust of the basin normal oceanic crust, and if so how thick is it? Otherwise what is the nature of the crust, and how thick is it?

20. During the next hour the earth's rotation about its axis is to be brought gradually to a halt. Describe the consequent sequence of events referring especially to the situation after: a week, a year, 1000 years, a million years, 100 million years.

21. Criticize the assertion that 'once isostatic equilibrium is achieved all the forces on a slab are in equilibrium'.

22. Comment on the assertion that the internal effects of the temperature variation over the earth's surface are restricted largely to the upper 100 km and will be manifest solely as a geoidal bulging of amplitude about 50 m.

23. We place a 35 km thick slab of continental material of density 2·7 g cm^{-3} on a *dry* crustless earth of mantle density 3·3 g cm^{-3}. (*a*) How much sticks out of the mantle?

(*b*) Water is now deposited on the surface of the earth covering the mantle to a depth of 2 km. Estimate the consequent vertical displacement of the continent. (*c*) Vigorous erosion sets in removing continental material at a rate averaging 2 km of thickness per 10^7 years, until the upper surface of the continent is at sea-level. How long will that take?

24. A mechanical spring is connected across a dashpot. One end of the combination is held in a fixed position; the other end is free to move. On its own the spring extends 1 cm for a force of 100 g-wt. On its own the dashpot extends at a rate of $1 \mathrm{cm s^{-1}}$ for a force of 30 g-wt. The free end of the combination is suddenly pulled with a force of 1 kg-wt. Estimate the maximum velocity of the free end during the subsequent displacement and the time required for half the final displacement.

25. Specify the parameters of a parallel mechanical spring, dashpot, and mass combination which simulates the earth's free vibrations 100 times faster and for which the amplitude of a vibration falls to 0·1 of its initial value in 10 cycles.

26. A jelly has been allowed to set in a cylindrical mould 20 cm in diameter and 1 m long. The mould is then upended on to a horizontal surface and quickly removed. Describe what happens and why.

27. Determine the apparent viscosity of granular material by measuring the quantity discharged in a given time through a vertical pipe fed from a funnel full of the material, using the relations: kinematic viscosity $v = \pi g \xi/128$, $\xi = d^4 p'/q$, where d is the inner diameter of the pipe, q is the volume discharged per unit time and $p' = h/l$, where h is the height of reservoir surface above the tube inlet and l is the length of pipe. With pipes of diameter 0·1–1·0 cm and, say, table salt of grain diameter about 0·3 mm, you should find $v \sim 1 \mathrm{cm^2 s^{-1}}$. Repeat your measurements with a variety of grain sizes and discharge rates. Comment on your results. To what extent can a granular material be regarded as behaving like a viscous material? During an ash eruption of a volcano what will be the effective viscosity of the material in the vent?

28. Estimate the total internal loss of energy of the earth given that the global average is $50 \mathrm{mW m^{-2}}$, and hence find the equivalent mass rate in $\mathrm{g s^{-1}}$ if this amount of power was obtained by converting hydrogen into helium. Assuming this rate was maintained through geological time how many grammes of hydrogen would be needed?

29. How long will it take to cook an ox? *Hint.* Obtain data on the cooking time and physical dimensions of commoner items such as peas, eggs, and potatoes, and plot x^2 against time t, where x is the narrowest dimension. Explain this relationship.
  Hence estimate how long it would take to cook the earth.

30. Given only the following information estimate and plot the mean interior temperature as a function of depth to a depth of 1000 km. (*i*) A wet granite melts at 750 °C, at surface pressure. (*ii*) A dry basalt melts at 1100 °C, at surface pressure. (*iii*) A clinopyroxene found in kimberlite melts at 1450 °C and 20 kbar. (*iv*) Mantle material has a coefficient of cubical expansion 10^{-5} per K, and specific heat $1 \mathrm{kJ kg^{-1} K^{-1}}$. Given that diamonds require a temperature of 2000 °C and a pressure of 30 kbar deduce the amplitude of the temperature pulse required to trigger a kimberlite eruption.

31. A hot liquid earth is at 3000 K. Assuming that the interior is well mixed and that the atmosphere has a transparency of 50 per cent estimate how long before the surface temperature has fallen for the formation of the first solid: (*a*) basalt; (*b*) granite.

32. The earth is to be blown up. (*a*) Estimate the amount of energy required. (*b*) Is there a suitable energy source available? (*c*) Assuming that the material produced is a perfect gas of mean molecular weight 20 make an estimate of the temperature at the centre of the gas cloud.

33. Calculate approximately the change in radius of a globe of mass 10^{22} kg as it freezes, assuming that the matter is incompressible and that from an original liquid of density $6.2 \, \text{g cm}^{-3}$ a solid mantle forms of density $4.1 \, \text{g cm}^3$.

34. Show that the radius r of a mass M of compressible material is approximately given by $\frac{4}{3}\pi r^3 \rho_0 / M = 1 - CGr^2/3\rho_0^{n-2}$. What would be the radius of an entirely basaltic earth?

35. Big G just increased by 10 per cent. What is the new radius of the earth? What else happened?

36. Why are there 365·242 days in a year?

37. Estimate the efficiency of the global machine where efficiency is taken to be the power used in rearranging the earth as a fraction of the total power used.

38. As in the egg model, assume that the earth is made of a mantle and core of in-compressible material. If the radius of the core has diminished uniformly with time estimate the gravitational energy released after 1000, 2000, 3000, 4000, 5000 million years. If the initial temperature of the earth was 3000 °C, the surface heat flux was a constant $120 \, \text{mW m}^{-2}$, and the only energy sources are thermal and gravitational, find the interior temperature at the above times. Repeat your calculations for: an asteroid, the moon, Jupiter.

39. When will the earth no longer have a fluid core?

40. A solution of sodium chloride, copper sulphate, and chrome alum is made by dissolving 200 g of each of the constituents in 3 litre of warm water. The solution is placed in a beaker of diameter 15 cm, of which the sides and bottom, but not the top, are insulated (crumpled newspaper in a cardboard box will do); the whole lot placed in the deep-freeze compartment of a refrigerator. Describe what happens. Relate your observations to a geological phenomenon.

41. Make an order-of-magnitude estimate of the amplitude of the density fluctuations in the earth using the following data and assumptions:
  (*a*) The non-hydrostatic contribution to the moment of inertia is less than 10^{-5} of the total;
  (*b*) this contribution is produced entirely from n density anomalies, of similar horizontal and vertical length scale, located entirely in the upper mantle;
  (*c*) the net effect of these anomalies is proportional to $1/n^{\frac{1}{2}}$.

Is your estimate compatible with other data? Improve the model by further assuming that the anomalies are produced solely in a thermal sublayer and thereby estimate an upper bound for the thickness of the sublayer. If the sublayer is actually about 100 km thick, obtain a precise estimate of the probable value of the non-hydrostatic contribution to the moment of inertia, and comment critically on the modern measurements.

42. A hot metal object of mass 1 kg is totally immersed in a stream of a cold viscous fluid. Its temperature falls to half of its initial temperature in an hour. The measurement is repeated with a body of similar shape but of mass 30 kg. How long will it take to lose half its temperature?

43. An approximately spherical mass of diameter 100 km which maintains its relative density contrast $\Delta\rho/\rho = 0.05$ is released at the top of the mantle. Assume that the kinematic viscosity of the mantle is $10^{16}\,m^2\,s^{-1}$. How long will it take to reach the core.

44. Outline the design of an unmanned vehicle for travelling through the mantle. Mention the measurements to be made and describe the telemetry system to be used. Estimate the cost of your system.

45. Lord Rayleigh observed, when cooking porridge, that a thin layer was prone to stick to the base of the pan and burn. Given that the minimum safe depth of porridge is 0.05 m and the critical Rayleigh number is of order 10^3, estimate the kinematic viscosity of porridge.

46. An ocean-floor nuclear reactor is to be cooled by placing it inside a sealed spherical container filled with sea-water. The reactor produces 100 MW of waste heat. What is the diameter of the container?

47. Estimate the thickness and time-scale of the thermal sublayer for a body of given temperature excess immersed in the following fluids: (a) $100\,K$, air; (b) $10\,K$, water; (c) $1200\,K$, liquid basalt; (d) $1000\,K$, solid basalt.

48. Estimate the thickness and temperature-difference across the thermal boundary layer in the core at the core–mantle interface. Hence estimate the amplitude and lifetime of the density fluctuations in the core.

49. What is the temperature at the centre of the earth?

50. Obtain a global estimate for the viscosity of the earth assuming: its initial mean temperature was 4000 °C and is now 2000 °C; the surface heat flux after 5×10^9 is $50\,mW\,m^{-2}$.

51. Investigate possible thermal histories with the following simple model.

A planet has: (*i*) an interior which is well mixed and has a uniform temperature; (*ii*) a thin uniform outer layer through which heat is transmitted solely by thermal conduction and is sufficiently thin to be in a near steady state; (*iii*) a conductive layer thickness which is independent of time.

Calibrate your model with initial temperature 3000 K and surface heat flux $50\,mW\,m^{-2}$ after 5000 million years.

Estimate: (a) the surface heat flux after 1000 million years; (b) the present interior temperature of the moon. (c) the surface heat flux of Jupiter. Comment on your model.

52. Discuss the aspects of Archean geology that could provide data to calibrate thermal-history models.

53. Describe what palaeontological evidence you would look for to support or contradict the hypothesis: during early geological history the earth's surface was completely inundated; this was followed by an intermediate period of temporary localized land areas, and that the subsequent emergence of more or less permanent land areas occurred at a fairly distinct time. If the hypothesis is correct thereby estimate the times of first appearance of temporary land and permanent land. Suggest a palaeontological project designed to calibrate thermal histories.

54. Redraw Fig. 5.5 (p. 36) to show the masses in the reactor vats at the time 500 million years after the beginning of geological time.

55. Discuss the comment that in describing the dynamics of planetary interiors on a large scale we have low power 'densities' but smaller systems have high power 'densities'.

56. Describe the geology of an asteroid.

57. A time-machine has been made. You were appointed as geologist for an expedition to 10 000 million years into the future. Describe what you expected to find. On your return you gave a half-hour résumé of your observations on TV. Give an outline of your lecture.

58. *The drift game.* Construct and use the following apparatus to simulate the gross rearrangement of the earth's crust. Alter the rules to make the effect as realistic as possible.
  Apparatus:

  (i) *The mantle.* A piece of peg board 1 m × 1 m—or a sheet of paper with a net of horizontal and vertical lines spaced 2·5 cm apart. A 10-cm strip on two opposite margins painted white—the polar regions.
  (ii) *Sediment.* A bag of marbles or small stones or small coins.
  (iii) *Continents.* Small, square or rectangular, cardboard trays about 5 cm × 5 cm.
  (iv) *Miscellaneous equipment.* 2 ordinary dice, a dice—the sign dice—with sides labelled only + or −. Set of paper clips and a set of labels.

  Rules:

  (i) *Start.* In turn each player places three boxes, each with a label in it, anywhere on the board. Boxes should always lie with their sides parallel to the lines and evenly over a set of lines. Place a marble in every uncovered hole and 10 in each box.
  (ii) *Play.* This involves for each round, about 20 million years, a turn for each player. A turn involves collecting or losing sediment and moving your continent after three rolls of the dice as follows.
    (a) Erosion: roll an ordinary dice. Remove from any one of your continents that number of marbles. If a box is empty it is removed from the board. The removed marbles are to be placed anywhere on the board.
    (b) East drift: roll the sign dice and an ordinary dice together. Move any one of your continents that number of lines east, or west if the sign is negative. Place in the box all marbles passed over.
    (c) North drift: similar to east drift.
    If your box will pass over the edge of the board, the move is reversed.
    If your box crashes into another box, it becomes yours provided the incident box holds 10 or more marbles—change the labels and clip the boxes together in a cluster. Following a crash any remaining displacement is ignored.
    If at any time immediately following drift a single box holds more than 20 marbles put 10 of them in extra box, taken from those previously removed, if there is one available, and place it anywhere. If the box is part of a cluster, just unclip it. Whenever 5 or more boxes or parts of boxes are in the polar region there is an ice age. Increase the erosion rate threefold.
  (iii) *Finish.* Play for 1000 million years. The winner has the most marbles (or coins)

59. Two crustal slabs A and B, with A lying to the west of B, sit on a mantle of heat flux 50 mW m^{-2}. They are 20 km thick of mean elevation 100 m above sea-level and of thermal conductivity 2 W m^{-1} K^{-1}, except that slab A is 2000 km wide and a 500 km wide region on its western margin has thermal conductivity 1·4 W m^{-1} K^{-1}; slab B is 1000 km wide and has a similar region of low thermal conductivity on its eastern margin. Between the slabs there is 1000 km of oceanic crust on which there is a 1 km layer of sediments of density 2·4 g m^{-3} and thermal conductivity 1·7 W m^{-1} K^{-1}.

(a) What is the speed of approach of the two slabs?

(b) Assuming that the sediments accumulate in a pile of rectangular section in isostatic equilibrium estimate the time elapsed when the relative speed of A, B is zero.

(c) What is the direction and speed relative to the mantle of the entire slab, A + sediment pile + B, at that moment?

(d) Orogeny takes place and the bulk of the sediments are metamorphosed to granites of density 2·7 g cm^{-3} and thermal conductivity 3 W m^{-1} K^{-1}. What is the relative speed of the whole slab at the end of the granitic phase?

(e) Erosion of the mountains lowers the surface to the same level as the surroundings. What is the relative speed of the whole slab then?

60. A nearly circular crustal slab is rotating at a uniform rate about a pole at its centre. Assuming that the slab is rigid with finite strength 0·1 kbar and floats on a mantle of kinematic viscosity 10^{17} m^2 s^{-1}, deduce the diameter of the slab. Discuss the factors which determine the smallest possible independent crustal slab.

Why are continents the shape they are?

61. A thick slab of granitic material floats at rest on a turbulent mantle of given mean temperature. Discuss the processes involved in melting off the lower part of the slab. Deduce a relation between the interior temperature and the thickness of the slab when melting ceases.

Discuss why the granitic parts of the crust are so persistent.

62. Why does the solid surface elevation lie between −10·96 km and 8·79 km? Why not between −20 km and +30 km? What would the range of surface elevation be 4000 million years B.P.?

63. Why is the oceanic crust only 5 km thick.

64. Since the beginning of geological time how many ocean basins have been created and destroyed.

65. How many super-continents, of area greater than 20 per cent of the earth's surface, have there been since the beginning of geological time.

66. Estimate the permeability of the upper mantle beneath an active oceanic ridge of height 4 km. Assume that: the magma percolates up a uniform permeable channel of depth 100 km and width 10 km; the pressure difference driving the magma is the difference between the lithostatic and hydrostatic head; new oceanic crust of 5 km thickness is formed at a rate of 30 km per 10^6 yr. Sometime later the oceanic spreading rate falls to 10 km per 10^6 yr. What is the new height of the ridge?

67. A long sedimentary basin 1000 km wide and 3 km deep is laterally compressed to a width of 30 km. Estimate the energy per gram required.

68. Suppose the earth had been built without a hydrosphere. Discuss what difference that would make to the dynamics of the upper mantle and crust.

69. Describe the crustal geology of Jupiter.

70. Compare an orogenic system to the extrusion of matter from a press with non-deformable walls. Discuss the cases of: (a) a purely viscous material, and (b) a material of finite strength. Has the model any validity?

71. A rigid rectangular continental slab of density $2.7\,\mathrm{g\,cm^{-3}}$ is floating in the mantle of density $3.3\,\mathrm{g\,cm^{-3}}$. The slab is of uniform thickness $30\,\mathrm{km}$ and of width of $2000\,\mathrm{km}$, except that there are two mountain ranges. The first, $4\,\mathrm{km}$ high with a corresponding root of $18\,\mathrm{km}$ (giving a total thickness of crust of $52\,\mathrm{km}$), is $200\,\mathrm{km}$ wide and located parallel to an edge between 500–$700\,\mathrm{km}$ from the edge; the second, $2\,\mathrm{km}$ high with a $9\,\mathrm{km}$ root, is parallel to the same edge and lies between $1500\,\mathrm{km}$ and $1900\,\mathrm{km}$ from the same edge.

(*a*) Describe the hydrostatic forces acting on the slab and plot a graph of vertical stress at the level of the ambient mantle as a function of distance from the edge.

(*b*) Erosion removes the $2\,\mathrm{km}$ mountain range. Describe the new stress distribution.

(*c*) Suggest what else might happen in both situations.

72. Discuss the factors which determine the transverse profile of an orogenic system. Consider the two types: continental margin (Andes) and intercontinental (Himalayas).

73. Sketch a graph showing the total mass, as a function of time, in an orogenic system. Indicate appropriate orders of magnitude.

74. Estimate the mass of rock-substance being processed in orogenic systems now. Thence make an estimate for 4000 million years B.P.

75. How often, on average, will there be on earth a mountain range with peaks reaching greater than $6\,\mathrm{km}$? Assume that 30 per cent of the earth's surface is covered by a relatively permanent granitic crust, erosion is extremely rapid, and crustal displacements average 3 cm per year.

76. Try to relate the observed temporal and spatial differences between the mio- and eugeosynclinal regions solely to the hypothesis that the former is made of older thicker sediments.

77. Compare the rocks and structures to be found in two old orogenic systems which differed only in the rate of sedimentary accumulation in the original geosyncline: one at the rate $5\,\mathrm{km}$ per $10^6\,\mathrm{yr}$, the other $0.5\,\mathrm{km}$ per $10^6\,\mathrm{yr}$.

78. A ball-bearing falls $10\,\mathrm{cm}$ through glycerine in about $25\,\mathrm{s}$. Hence, given the following data, estimate the time for a granitic pluton to rise $40\,\mathrm{km}$. Diameter of ball $1\,\mathrm{mm}$, diameter of pluton $10\,\mathrm{km}$; kinematic viscosity of glycerine $10^{-3}\,\mathrm{m^2\,s^{-1}}$, kinematic viscosity of the granitic crust $10^{19}\,\mathrm{m^2\,s^{-1}}$, density ratio $\Delta\rho/\rho$ for the ball 7, for the pluton -0.05. Comment on your result.

79. A cumulus cloud is a large organized lump of air, heat, water, and electricity. What are the analogous ingredients of a large igneous intrusion.

80. Perform the following experiment and describe in detail your observations. Make a thin flour paste by first taking a tablespoon of flour and mixing a little water at a time until the paste is smooth. Top up with two cups of water and stir well. Place in a pot of 10–$15\,\mathrm{cm}$ diameter on the stove, at first with a very low heating rate. At each stage measure the fluid temperature. Slowly increase the temperature until a surface skin forms and foaming commences.

Remove the pot from the heat and place the pot so that the sides and base are insulated—crumpled newspaper is suitable. Record the temperature and time. Explain your observations.

What geological phenomenon could be modelled in this way? Relate in detail your observations to what is known of the corresponding geological phenomenon.

81. An ultrabasic sill of thickness 100 m is injected beneath a shallow sedimentary layer. Owing to cooling through the top of the sill a temperature difference of 50 K is established across the sill before it begins to freeze. The coefficient of cubical expansion of the magma is $2 \times 10^{-5} \text{ K}^{-1}$. The magma contains 20 per cent of a heavy constituent which contributes to the magma density 50 per cent more than the other constituents. Deduce the spacing between the layers which form and make a rough estimate of the thickness of the transition zone at each interface. Will the sill have negligible or a large remnant magnetism?

82. You have been invited to give a lecture on volcanism to a young audience at the Royal Institution, London. Describe the demonstrations and apparatus you will use.

83. In the year A.D. 2500 there will be a great shortage of building material. It is therefore proposed to start controlled ignimbrite eruptions by releasing energy at 30 km depth at the base of the continental crust. Estimate the amount of energy required and the total mass output.

84. Verify the statements made in Chapter 14 about reservoir clocks. Obtain some cans of various diameter and some tubes of various but small internal diameter and length. Measure the capacitance C and resistance R of the cans and the tubes. Fit the tubes in turn to the cans and measure the time taken for a can to half empty. Plot time against RC.

Connect three different cans all sitting on a flat surface in a ring with three equal tubes, the output of one feeding the input of the next, and at time zero fill up one can. Measure the elevation of fluid in each can as a function of time. Plot a graph of your results and explain it. In particular, discuss the reason for the more elaborate motion in the smallest diameter can and identify the time scale of the combined system.

85. Determine the area of the base of a volcanic island as a function of time assuming that its shape is determined by the condition of finite strength, and the magma source in the mantle is constant.

86. Estimate the age of a volcanic island which rises 10 km above the ocean floor and erupts 0.1 km^3 every 10 years.

87. Estimate roughly the total number of eruptions which have occurred on the island of Hawaii.

88. Assume that the diameter of the vent of a central volcano is determined by the condition that during the peak output the hydrodynamic drag balances the finite strength of the wall rock. Hence, using data from known volcanoes, show that the drag coefficient λ is about 0.01. Then predict the vent diameter for a volcano of peak output 10^{11} ton per day. Has there ever been such a monster?

89. Many volcanic vents are trumpet-shaped. Assume this arises because in the top of the vent the hydrodynamic wall stress exceeds the finite strength of the wall rock, thereby stripping rock off and widening the vent until the stress is below the finite strength. Estimate the vent shape for a volcano that at peak output discharges 3×10^8 ton per day from a reservoir of a wet basaltic melt of density 2.7 g cm^{-3} of which 10 per cent by mass is water when the wall rock has a finite strength of 1 kbar. *Hint.* First show that $C^2 \approx \lambda m^2 / 4\rho\tau$ and simply find $p(z)$. Relate your results to the shape of volcanic pipes.

90. In the derivation of the age of a lava pile, apart from the evaluation of the working head, the quoted formula takes the density of the ambient rock and the working fluid to be the same. Derive the complete expression for the age and show that the error involved in the simplification is negligible in practice.

91. Estimate the rate of discharge from a vertical dyke of width 100 m, length 10 km, and depth 10 km, assuming it is fed from a reservoir of negligible internal resistance at a depth of 50 km. The discharge spreads over an area of 300 km^2 to an average depth of 30 m. How long did the flow last? Why did it stop? After what interval of time could a similar eruption occur again? Eruptions continue in that region for 1 million years. What is the thickness of the lava plateau?

92. A hollow container is made from two large vertical cylinders joined by two thin horizontal pipes of length l, internal radius r, a vertical distance h apart. The container is filled with a fluid of density ρ, kinematic viscosity v, coefficient of cubical expansion γ. The fluid in one cylinder is maintained by thermostats at a temperature θ higher than the temperature of the other. Obtain an expression for the volume of fluid per unit time, Q passing through one of the pipes. For the case $h = 10$ km; $l = 2$ km, $r = 10$ m; $\rho = 2\cdot7$ g cm^{-3}; $\theta = 200$ K; $v = 0\cdot1$ m^2 s^{-1}; $\gamma = 3 \times 10^{-5}$ K^{-1}, find: (*a*) Q; (*b*) the power transferred between the vertical cylinders; (*c*) the power dissipated as viscous work in the pipes; (*d*) the efficiency of the system; (*e*) the time taken for the temperature of the hot cylinder of fluid to fall $0\cdot2\theta$, if the thermostats are removed and the system is insulated.
  Comment on a geological application of your results.

93. Discuss fluidization in geological systems, paying particular attention to possible global fluidization during initial degassing of the interior, and the role in lithothermal systems and in volcanic eruptions.

94. You have just been awarded a grant of £10^9 for a research project, the results of which are to be used in 5 years time to plan a prototype engineering project for direct exploitation of the mantle at a depth of 100 km. Describe the studies to be undertaken in your project.

95. If all the heat flow through the earth's surface arose from steaming ground estimate the mass discharge per second. What proportion of the earth's surface area would be required if this discharge was from ground with a very shallow water table at 80 °C?

96. Estimate the time required for a volcano to be cooled after an eruption by the free circulation of water in a permeable water saturated ambient crust. Use the following data: depth of permeable upper crust 10 km; permeability 0·1 darcy; vent reservoir a vertical cylinder of diameter 0·1 km. Compare this time with the corresponding time for no convection of water. The island of Oshima, Japan, after an eruption, as revealed by changes in its magnetism, cools below the Curie point in about a year. Make a rough estimate of the permeability of the island.

97. Bloggs has to advise on the possibility of constructing a 100 MW geothermal power station. This is all the information he has: (*i*) over an area of 10 km^2 there are some small hot pools; (*ii*) the visible rock is rather homogeneous with fine joints of width about 0·1 mm roughly every 1 m. What should be Bloggs's advice?

98. Estimate the two quantities: (*a*) the total energy stored in the Taupo hydrothermal systems; (*b*) the total energy stored in a similar crustal block of normal temperatures. Why is (*a*) so much smaller than (*b*)?

99. Write an essay on 'Identification of geological processes by measurement of a minimum number of key parameters'. Base your essay on the discussion of the following.

(*i*) What is the least number of questions of the type 'Male?, shorter than 150 cm?,.,. etc,' for which the only allowed answers are 'Yes or Otherwise' (which includes: no; don't know; not applicable; etc.) required to uniquely identify everyone on earth. Prepare an outline of a possible list of questions.

(*ii*) Rock-types vary continuously but it is convenient to classify rocks by means of certain criteria based on composition, etc. Assume this has been done to put all but the rarest of rocks into 1000 types. In practice only about 30 of these types are frequently encountered. We wish to prepare a key for identification of rocks in hand-specimen in two parts: the first to identify the 30 commonest plus the single entry 'others', the second to identify the 'others'. Identification is to be by questions answered 'Yes or Otherwise', as in (*i*). How many questions are in each of the parts of the key? What are the questions in the first part of the key?

(*iii*) Make a list of the key parameters used in this book. Extend the list with parameters of your own choice to a total of 20. Order your list on the certainty with which the parameters are known or if unknown the probability that they could be measured in your lifetime.

100. For the next 10 years all the scientists currently studying the earth are to work solely on five topics of your choice. What are they?

Glossary–Index

This glossary–index includes definitions or brief explanations of technical terms and physical concepts, definitions of symbols, data items for frequently used values, frequently used simple formulae, and references to authors or articles other than those in the Reading List. Each item for which there is a symbol is entered twice: by name—which gives the symbol; and by symbol—which gives the name, and where appropriate, other information. Greek-letter symbols are listed in the place corresponding to the English spelling of the name of the letter. Where symbols are used temporarily in one section only and are fully defined there, they are not in this glossary–index.

The quoted numerical values should be treated with an appropriate degree of scepticism. There are many pitfalls. Some quantities are accurately known, some to order of magnitude, and several are guesswork. Quantities may be quoted in the literature with a variety of values even in the same book. The greatest bane are quantities which are traditional, having been stated years ago, and because they have been quoted and requoted many times are regarded as correct. The accuracy taken for granted in physics is not a feature of geology. There are many aspects of geology for which we as yet literally do not know which way is up.

Page numbers are given in square brackets at the end of the index entries, or occasionally in the middle of an entry at the end of an appropriate sub-section.

a Present radius of earth, generally mean radius 6371 km, equatorial radius.

A Principal moment of inertia about equatorial axis; area.

abundance So-called cosmic, relative to
Si = 1; H, 2.6×10^4; O, 23.6; C, 13.5; N, 2.4; Mg, 1.0; Fe, 0.9; S, 0.5; Al, 0.09; Ca, 0.07; Na, 0.06; Ni, 0.05; P, 0.01.

abyssal plain Very flat portions of deep ocean basins, slope 1:1000; oozes interbedded with turbidity current deposits.

acidic Rock rich in equivalent silica, often with free quartz present. Typical example: granite. *See also* igneous rock.

adiabatic gradient A well-mixed system, outside thermal boundary layers, is approximately isentropic if averages are taken over a suitable time-interval, when the temperature gradient is $-dT/dr = \gamma g T/c$, where γ is the coefficient of cubical expansion, c the specific heat at constant pressure. About 0.2 K km^{-1} on earth. [32]

advection Refers to the transport of a quantity by the motion of the material in which it is contained. In the conservation equation for the quantity θ the advection term is $q \cdot \nabla\theta$ per unit volume. A convective system is one driven by buoyancy forces and in which advection of density is a dominant transport mechanism. [62]

Africa [120, Table 12.2]

age Time elapsed since some distinct short-lived event as measured by the change in some distinct component. In geology, as in other subjects, placed largely outside historical time, there is no time-scale even remotely as absolute or as consistent as present-day civil time. Fortunately there are many natural phenomena which proceed more-or-less monotonically with frequent distinct markers which allow long time-sequences to be established. For time-intervals spanning geological time radioactive methods are currently the best, but they rely on careful calibration, careful selection of samples, and very strong assumptions about conditions in the past. The commonly quoted age of the earth is based on the proportions of isotopes of lead ^{206}Pb, ^{207}Pb, and uranium ^{235}U, ^{238}U (*see* radioactive elements) using $t = \{\ln(1 + p/u)\}/\lambda$, where t is the age, p is the proportion of lead atoms, u is the proportion of uranium atoms, and λ is the decay constant.

Simple model. Both lead isotopes are assumed to be of zero initial concentration. For the nuclear reaction $^{238}U \rightarrow {}^{206}Pb$ with $p:u = 1\cdot89:1$ now, and $\lambda = 0\cdot154 \times 10^{-9}\,yr^{-1}$ we find $t = 6\cdot9 \times 10^9\,yr$. For the nuclear reaction $^{235}U \rightarrow {}^{207}Pb$ with $p:u = 217:1$ now, and $\lambda = 0\cdot98 \times 10^{-9}\,yr^{-1}$ we find $t = 5\cdot6 \times 10^9\,yr$. Considering the extreme assumptions, these two ages are quite consistent.

Second model. As before we assume that the two lead isotopes are initially of the same atomic proportions, but not zero. We then have two decay equations involving the quantities: t and the initial amount of ^{206}Pb and the initial amount of ^{207}Pb. We find the proportion of initial ^{206}Pb relative to its value now to be $0\cdot4$; the proportion of initial ^{207}Pb relative to its value now to be $0\cdot5$ and the age $t = 4\cdot8 \times 10^9\,yr$. This method of improving the age model is to be preferred to the more usual use of data from meteorites (*see also*). All these and similar estimates should be treated with extreme scepticism.

age of earth 5×10^9 yr. [5, 42]

age of ocean floors *See* ocean floor.

air Density at $0\,^\circ C$ and 1 bar pressure: $1\cdot19 \times 10^{-3}\,g\,cm^{-3}$; velocity of sound, $330\,m\,s^{-1}$.

alkalic Refers to presence of one or more of the elements Li, Na, K, Rb, Cs, Ca, Sr, and Ba.

α Compressional wave velocity.

America, North [120, Table 12.2]. Polar wandering curve [86, Fig. 10.1]

America, South [144]. *See also* Andes, Chile.

amplitude ζ

Andes [108, 116, Fig. 12.1; 117, Fig. 12.2]

andesite Originally named after a series of lavas in the Andes, a large group of rocks of compositions lying between granite and basalt, extruded from within the continental crust and its margin. Normally a fine-grained mixture of plagioclase feldspar and pyroxene or biotite.

andesitic volcanism [148, Fig. 14.4; 149]

anelasticity [24]

angle θ

ångström Unit of length, 10^{-10} m. Roughly the spacing of atoms in a crystalline solid.

angular velocity ω. Of earth, $7\cdot29212 \times 10^{-5}\,rad\,s^{-1}$, daily variation of rotation period 1–2 ms.

archean A somewhat vague expression, originally intended to refer to the time before the Proterozoic. Here merely used in the sense of times before very roughly about 1000 million years B.P.

ash eruption [163]

Asia [151]

Aso-san caldera, Japan [144, Fig. 14.1]

assimilation Complete incorporation of host material by magma.

astronomical unit $1\cdot496 \times 10^8$ km, the sun's mean distance from the earth for a parallax of $8\cdot80''$.

astrophysics SMITH, E. and JACOBS, K. (1973). *Introductory astronomy and astrophysics*. W. B. Saunders, Philadelphia.

Atlantic, south [108]

atmosphere Mass 5×10^{18} kg; scale height of equivalent homogeneous atmosphere 8 km; density at sea-level, $1\cdot3 \times 10^{-3}\,g\,cm^{-3}$. Pregeological [41]

atm (atmosphere) Unit of pressure, the nominal pressure at the base of the atmosphere $= 1\cdot01325$ bar.

atom Typical size 10^{-10} m, mass $1\cdot66 \times 10^{-27}$ kg per unit atomic weight, cf. mass of electron $9\cdot1 \times 10^{-31}$ kg.

atomic mass unit (a.m.u.) Mass of atom of unit atomic weight based on $^{12}C = 12$ a.m.u.; 1 a.m.u. $= 1\cdot660 \times 10^{-27}$ kg.

atomic weight, mean for elements on earth 22, estimated from the empirical relationship $\rho(P)$ as a function of mean atomic weight, originally after Birch.

Australia, polar wandering curve [86, Fig. 10.1]

average quantities [65, 70]

Avogadro's number 6.023×10^{23} mol^{-1}.

b Parameter of viscosity variation with temperature, nominally 10^4 K [27, 81]

B A principle moment of inertia about an equatorial axis.

bar Unit of pressure or stress, 10^5 N m^{-2} = 10^6 dyn cm^2. About 1 atm, or the pressure under 10 m of water or 3 m of rock.

basalt Basic igneous rock, usually extrusive, containing calcic plagioclase feldspars, pyroxene, and often olivine. The most abundant igneous rock and lava. Broadly divided into two main groups. Alkali basalts: of low equivalent silica, always with olivine, found in oceanic basins and believed to be produced at depths down to 100 km. Calcalkali basalts (which include tholeiites—basalts with calcium-poor pyroxenes): a wide range of compositions from picritic (olivine-rich) to andesitic basalts found in association with increasingly siliceous volcanics up to rhyolites, in mountain belts, lava plateaux, the lower lavas of oceanic islands such as Hawaii and believed to be produced at relatively shallow depths of about 20 km. [19, 32, 126; 146; Table 14.2]

basic Rock about half or less of which is equivalent silica. Typical example, basalt. *See also* igneous rock.

basic (simple) model [49]

basement General term for a regional undermass of metamorphic or igneous, or highly folded rock underlying a sedimentary or younger cover.

bell [11]

Benioff zone [105, Fig. 11.6; 107, Fig. 11.7]

β Shear wave velocity. Also concentration ratio.

bladder model The original example is due to Osborne Reynolds (*see also*), the extension to CHARLES FRANK (1965). On dilatancy in relation to seismic sources, *Rev. Geophys.* **3**, 485–503. [25]

biomass Total quantity of living matter, $\sim 10^{15}$ kg; global average ~ 1 kg m^{-2}; efficiency relative to total solar energy 0.1–4 per cent.

biosphere The oceans, lowest portion of atmosphere, and crust to depth of order 100 m.

boiling-point [175]. *See also* BPD.

Boltzmann constant 1.381×10^{-23} J K^{-1}.

Bouger anomaly Local gravity adjusted to a reference elevation, mean sea-level, by removing contributions due to ellipsoid of entire earth, elevation, and material between observation point and sea-level treated as a uniform infinite slab. Broadly indicates anomalous mass distribution below the reference level.

boundary-layer thickness δ. Upper-mantle scale, nominally 100 km.

Boussinesq Approximation in convection theory. [57, 60]

B.P. Before present.

BPD The boiling-point of water at a given depth; 120 °C at 10 m; 134 °C at 20 m; 160 °C at 50 m.

brittle fracture [28]

buoyancy [55, 60]

c Polar radius. Also general wave velocity. As subscript: c − core.

C Largest principal moment of inertia, that about polar axis: $A < B < C$. Elastic parameter = $\chi\rho^n$. Also concentration. Also drag coefficient.

caldera Extensive subsided volcanic crater. [144, Fig. 14.1]

calorie Unit of energy, 4.186 J; international calorie (now obsolete) roughly the energy required to increase the temperature of 1 g water by 1 K.

capacitance of a reservoir Change in volume of fluid in the reservoir per unit change in pressure. For a reservoir with a free surface and pressure measured as a head of reservoir fluid, the capacitance is the area of free surface. [153]

Caucasus, U.S.S.R. [120, Table 12.2]

centre of earth 6371 km down; density about $13 \, \text{g cm}^{-3}$; Young's modulus 3·2 Mbar; incompressibility 15 Mbar; Poisson's ratio 0·46 (cf. 0·5 for fluid); reached by *PK* wave in 10 min; temperature perhaps 3000 °C; pressure 3·67 Mbar.

centrifuge technique The method is that of Hans Ramberg. The studies of orogeny [125] and sills [152] are but a few of the very beautiful results of his work.

Chandler motion 0·1″ wobble of pole of rotation of periods 1 year and 14 months.

chemistry [37]

χ Compressibility.

Chile, 1960 earthquake spectrum [12, Fig. 3.2]

circulation [103]

circulation of water in crust [166]

clock, volcanic [153]

cloud in space [41]

cold water in hydrothermal system [171]

compressibility $\chi = \dfrac{1}{\rho}\left(\dfrac{\partial \rho}{\partial P}\right)$. Typical values in the earth are of order 0·1 Mbar^{-1}. [15; 17, Fig. 3.5]

compressional wave [13]

concentration of elements [45]

conduction *See* thermal conduction.

constants CLARKE, S. P. (ed.) (1966). *Handbook of physical constants.* Geological Society of America.

construction of the earth [Chapter 6]

continent A symposium on continental drift. *Phil. Trans. R. Soc.* A**258** (1965).
  disruption [96, Fig. 11.1; 109, Fig. 11.9, 110, Fig. 11.10]
  drift [87, 100, 106]
  lower crust [19]
  margin [18; 99, Fig. 11.2]
  migration [87]

convection A heat-transfer system involving gross displacement of matter. *See also* advection.
  cellular [53]
  forced [51; 52, Fig. 7.1; 111]
  free [52; 51, 53, Fig. 7.2]
  models [79, Table 9.1; 82, Table 9.2]
  penetrative [131, 139]·
  permeable (or porous) medium [129; 130, Fig. 13.1]
  problems [Questions 42, 43, 45, 47, 92]
  turbulent *See* thermal turbulence
  vigorous [58]
  water in the crust [166]
  weak [54]

cook Elevate the temperature of a body to some definite temperature by heating the outside and transferring heat solely through the body by thermal conduction. The time required is proportional to h^2/κ, where h is a typical dimension, in practice the smallest thickness, and is not a function of the temperature of the oven. [122; Question 29]

coordinates Cartesian x, y, z.

core Major, largely fluid inner part of the earth, radius = 3470 km = 0·545a. Outer core depth, 2900–4980 km; transition zone, 4980–5120 km; inner core (presumed solid) 5120–6370 km.

Density range roughly $9 \cdot 5$–$13 \, \mathrm{g \, cm^{-3}}$. Mass, $1 \cdot 9 \times 10^{24} \, \mathrm{kg} = 0 \cdot 32 \times$ mass of earth. Viscosity, virtually unknown, perhaps $10^{-2} - 10^{10} \, \mathrm{P}$. Only the very simplest aspects of size, mass, and formation of the core are discussed in this book.

craton Extensive and relatively permanent portion of the granitic crust.

crust HART, P. J. (ed.) (1969). *The earth's crust and upper mantle.* American Geophysical Union Monograph, No. 13, American Geophysical Union, Washington D.C.

The crust consists of outer layers of earth with P wave velocities somewhat less than $8 \, \mathrm{km \, s^{-1}}$. Variable thickness typically 5 km oceanic; 30 km continental. Mass, $2 \cdot 5 \times 10^{22} \, \mathrm{kg} = 0 \cdot 004 \times$ mass of earth. Relative abundances of elements as number of atoms per 1000, given by Mason: O, 626; Si, 212; Al, 65; Fe, 19; Ca, 19; Na, 26; K, 14; Mg, 18. Owing to its relatively large atomic volume, oxygen represents about 94 per cent of the volume of the crust. Thus crustal rocks can be thought of as a mesh of oxygen (and silicon) atoms with a sprinkling of the other elements mentioned above, and an extremely small quantity of others.

crystal A homogeneous solid, generally of a definite geometrical form, with a highly regular atomic lattice. Primarily classified by composition and form in six systems of 32 possible symmetries: cubic, tetragonal, hexagonal, orthorhombic, monoclinic, triclinic.

crystal friction Effective viscosity $= 10^{16} \, \mathrm{P}$. Viscosity from internal rearrangements with individual crystals. Global values obtained by measurements of the Q (*see also* Q) of a vibration of period τ and related shear modulus S from: $Q = 2\pi\mu/S\tau$. $S \sim 1$ Mbar, $\tau \sim 10^2 \, \mathrm{s}$, $Q \sim 500$. Estimate gives an upper bound.

cubical expansion coefficient γ.

Curie point, ferromagnetic Temperature above which spontaneous magnetization is zero. [87]

Darcy's law [129]. *See* permeability.

decay constant λ. *See* radioactive elements.

decay time τ.

deformation of shape Rate [92]

degassing [144]

density Mass per unit volume, ρ. Typical values for rock in $\mathrm{g \, cm^{-3}}$: granite $2 \cdot 67$, granodiorite $2 \cdot 72$; syenite $2 \cdot 76$; quartz-diorite $2 \cdot 81$; diorite $2 \cdot 84$; gabbro $2 \cdot 98$; peridotite $3 \cdot 23$; dunite $3 \cdot 28$; eclogite $3 \cdot 39$.

depth h, z, H.

DGP The degassing pressure at a given depth. [158]

diamond [32]

diapir An upward-penetrating core of material from depth.

differentiation, magmatic Fractional separation of magma into dissimilar components.

diffusion [51, 57, 60].

diffusivity variation, effect on structure [139]

dilation Proportional change in volume.

dimensional analysis PANKHURST, R. C. (1964). *Dimensional analysis and scale factors.* Chapman and Hall, London.

The study of how quantities which specify a system change when the scales of the system change. A subject not only fundamental to quantitative science but the direct basis of simulation and model studies. The book by Pankhurst is a good introduction. *See also* scale.

dimensions, and masses of parts of earth [36, Table 5.1]

discharge of magma [147]

discharge of water [169]

discharge velocity U.

dislocations [26]

dolerite A coarse-grained rock of basaltic composition intruded within the crust.

Donbass, U.S.S.R. [120, Table 12.2]

drag F, ξ. *See* hydrodynamic drag.

drift *See* continent, drift.

'drunkard's walk' [94]

dunite A rock composed largely of olivine. [39, Table 5.2]

dyke A thin sheet-like intrusion normally cutting steeply across pre-existing structures.
  eruption [147, 156]. *See also* flank eruption.
  swarm [111]

dynamo [87]

dyne Cgs unit of force $= 10^{-5}$ newton.

e Flattening ratio, $(a-c)/a$. *See also* Maclaurin.

E Energy in general. Also Young's modulus.

earth
BOTT, M. H. P. (1971). *The interior of the earth.* Edward Arnold, London. (The bibliography is also very useful.)
GASKELL, T. F. (ed.) (1967). *The earth's mantle.* Academic Press, New York.
GUTENBERG, B. (1959). *Physics of the earth's interior.* Academic Press, New York.
JEFFREYS, H. (1959). *The earth* (4th edn.). Cambridge University Press, London.
STEINHART, J. S. and SMITH, T. J. (1966). *The earth beneath the continents.* American Geophysical Union Monograph, No. 10. American Geophysical Union, Washington, D.C.
TAKEUCHI, H. (1966). *Theory of the earth's interior.* Blaisdel, Waltham, Massachusetts.
FLUGGE, S. (ed.); BARTELS, J. (group ed.) (1956). *Handbuch der Physik*, Bd. XLVII (Geophysics I). Springer-Verlag, Berlin.

earth (Geodectic reference system 1967) Equatorial radius $= 6378 \cdot 160$ km; polar radius $= 6356 \cdot 775$ km; flattening factor $= 1/298 \cdot 25$; radius of sphere of equal volume $= 6371$ km; area of surface $= 5 \cdot 101 \times 10^8$ km^2; volume $= 1 \cdot 083 \times 10^{12}$ km^3; mass $= 5 \cdot 977 \times 10^{24}$ kg; mean density $= 5 \cdot 517$ g cm^{-3}; normal acceleration of gravity: at equator $= 9 \cdot 780318$ m s^{-2}, at the pole $= 9 \cdot 832177$ m s^{-2}; mean solar day $= 86400$ s; sidereal day $= 86164 \cdot 09$ s; velocity of rotation at equator $= 465 \cdot 12$ m s^{-1}; mean moment of inertia $= 8 \cdot 025 \times 10^{37}$ kg m^2.

earth, diagrammatic section [2, Fig. 1.1]

earth seen from space [4, Fig. 2.1]

earthquake As opposed to microseismic (*see* microseism) activity: relatively large and localized release of strain energy by structural rearrangement. [13, 28, 107]

Richter scale, outline − 3, Detectable with a suitable array of local seismic stations; 0, readily detectable locally, 3×10^5 J; 3, felt nearby, about 10^4 per year; 5, some local damage; 6, destructive locally, about 100 per year, all shallow; 7, major earthquake, recorded globally, about 10 per year; 8 +, rare, normally catastrophic, 10^{17} J.

magnitude $M = \log A + 1{\cdot}66 \log \Delta + 2$, where A is the maximum surface wave amplitude of 20 s period, and Δ is the epicentral distance in degrees. Empirically related energy E in joules: $\log E = (4{\cdot}4 + 1{\cdot}5M)$. Total energy: 10^{18} J per year.

eddy [51, 53, 68, 70, 113]

egg model [8, Fig. 2.3; 9, Fig. 2.4]

elastic body Idealized body for which the stress and strain are linearly related. In theoretical studies we use as the two constants of proportionality λ and μ, the Lamé parameters. Elasticity theory gives the relations: incompressibility $K = \lambda + \frac{2}{3}\mu$; Young's modulus $E = \mu(3\lambda + 2\mu)/(\lambda + \mu)$; Poisson's ratio $\sigma = \lambda/2(\lambda + \mu)$; compressional wave velocity $\rho\alpha^2 = \lambda + 2\mu$; shear wave velocity $\rho\beta^2 = \mu$.
Note also:
$$K/\rho = \alpha^2 - \tfrac{4}{3}\beta^2,$$
$$E = 3(1 - 2\sigma)K.$$
The fairly realistic approximation $\sigma = \frac{1}{3}$ is used throughout this book.

elastic modulus S, σ, χ, E.

elasticity LOVE, A. E. H. (1927). *The mathematical theory of elasticity.* Dover Edition (1944). [15; 16, Fig. 3.4; 17, 23, Table 3.1]. *See also* mechanical properties.

electron Mass $9{\cdot}109 \times 10^{-31}$ kg.

ellipsoid (International) *See* earth, for dimensions. [6, Fig. 2.2]

energy E, U.
  compressional [38]
  consumption [33]
  gravitational [37]
  kinetic [39]
  orogenic belt [33]
  thermal [38]
  problems [Questions 28, 37]
  stocks [40, Table 5.3]

enthalpy Of a single thermodynamic system, $H = U + PV$, where U is the internal energy, P is the pressure, and V the volume. [38, 175, 180]

enthalpy change $\mathscr{H}$.

epicentre Point on the surface vertically above an earthquake source.

ε Extinction coefficient (radioactive transfer). Also random function.

equation of state For a single substance the relation between the thermodynamic variables which define the thermodynamic state: density, pressure, and temperature. Thus $\rho = \rho(P, T)$. For a given rock-substance P is much the dominant variable, so that for many studies we can ignore T and take $\rho = \rho(P)$. The equation of state used in calculations done for this book is given by $\chi\rho^n = C$, where χ is the compressibility, $n = 3$, and $C = 28$ Mbar^{-1} in the mantle, 168 Mbar^{-1} in the core. [16; 17, Fig. 3.5; Questions 16, 33–35]

Eratosthenes [6]

erg Cgs unit of energy $= 10^{-7}$ J.

erosion Total run-off $= 4{\cdot}6 \times 10^{16}$ kg yr^{-1}; dissolved load $= 4 \times 10^{12}$ kg yr^{-1}; suspended load $= 1{\cdot}2 \times 10^{13}$ kg yr^{-1}. [118; 120, Table 12.1; 150]

eu-geosyncline *See* geosyncline.

Europe [88, Fig. 10.1(c)]

evaporation [173, 177]

expanding earth [46]

experiments, do-it-yourself [Questions 4, 25, 26, 27, 29, 39, 58, 80, 84]

explosion seismology [19]

explosive TNT, $4.2 \times 10^5 \, \mathrm{J\,kg^{-1}}$. Nuclear: normally expressed as TNT equivalent—thus 100 Megaton gives $4 \times 10^{17} \, \mathrm{J}$, comparable to a major earthquake.

extinction coefficient ε.

f Flux of heat. Also frequency.

F Force, drag.

fabric The texture, arrangement of the portions of a rock or structure.

Fennoscandia [20]

fingering [139, Fig. 13.9]

flank eruption [154, Figs 14.8–14.10]. *See also* dyke eruption.

flashing Partial or complete phase change of liquid water to water and steam within the fluid body. [173]

flattening, relative *e* [7]

floating crust *See* isostasy.

flow
   crystalline solids [26]
   regimes [50, 53]
   slope [26]

fluctuations [65]

fluid dynamics
PRANDTL, L. (1953). *Essentials of fluid dynamics.* Blackie, London.
LAMB, H. (1932). *Hydrodynamics.* Dover Edition (1945).
BATCHELOR, G. K. (1967). *An introduction to fluid dynamics.* Cambridge University Press.
STREETER, V. L. (ed.) (1961). *Handbook of fluid dynamics.* McGraw-Hill, New York.
SCHLICHTING, H. (1955). *Boundary layer theory.* Pergamon Press, London. [Chapter 7]. *See also* (*Nu*), (*Pe*), (*Pr*), (*Ra*), (*Re*), dimensional analysis.

fluidization Gross movement of the parts of a granular material due to the percolation of a fluid through the material. [Question 93]

flux of heat *f*.

flysch Mixed deposits produced by rapid erosion during active orogeny. Graywacke turbidities prominent in eu-geosynclinal troughs, often in association with thick volcanic deposits.

fold [109]

force *F*.

foreland In orogeny, the resistant or stable block towards which the sedimentary pile is shoved.

fossil magnetism [87]

free vibrations In the text only seismic vibrations are considered. Other modes of vibration are possible. For example, as shown by Kelvin (1863), if the earth is considered as an incompressible homogeneous fluid sphere, free surface waves of the solid earth have circular frequencies, ω, where $\omega^2 = \{2n(n-1)/(2n+1)\}g/a$, $n = 2, 3, \ldots$. The mode of lowest frequency, $n = 2$, is an ellipsoidal deformation with period about 5600 s. [11, Fig. 3.1; 12, Fig. 3.2; Questions 13–15, 24, 25]

freezing model [45]

frequency *f*.

fumarole [175; 176, Fig. 15.9]

g Acceleration due to gravity (the force on unit mass); $g' =$ vertical component of local gravity variation. At the surface of a spherical body of mass M and radius a, $g = GM/a^2$. For the earth this gross value is $g = 9.82\,\text{m}^2\,\text{s}^{-1}$. At the earth's surface and throughout the mantle nominally $10\,\text{m s}^{-1}$. The usual practical unit for surface survey work is the milligal (mgal) $= 10^{-3}\,\text{cm s}^{-2}$. Thus, on the surface $g \approx 10^6\,\text{mgal}$. [16, Fig. 3.4; Question 17]. *See also G*.

G Gravitational constant, $6.67 \times 10^{-11}\,\text{N m}^2\,\text{kg}^{-2}$. The gravitational force between two balls of basalt each of 140 m radius and just touching each other is $\sim 1\,\text{N}$. [7; Questions 5, 7–11]

gal Unit of acceleration, $1\,\text{cm s}^{-2}$. *See also* gravitational acceleration.

galaxy Radius $4 \times 10^{17}\,\text{km}$; 10^{11} stars; mass $4 \times 10^{41}\,\text{kg}$. Our solar system about $3 \times 10^{17}\,\text{km}$ from centre. Nearest other galaxy: Andromeda at $10^{19}\,\text{km}$.

Galileo [25]

γ Coefficient of cubical expansion, nominal value $10^{-5}\,\text{K}^{-1}$.

gas constant $8.3\,\text{J K}^{-1}\,\text{mol}^{-1}$.

geochemistry
WEDEPOHL, K. W. (ed.) (1969). *Handbook of geochemistry*. Springer-Verlag, Berlin.
FYFE, W. S. (1974). *Geochemistry* (OCS 16). Clarendon Press, Oxford.

geodesy The study of the size and shape and related quantities of the earth. [5]

geoid The gravity equipotential surface which is close to present-day mean sea-level. [19; 20, Fig. 3.7; 93, Fig. 10.4; Question 22]

geological column Normally the total thickness of sedimentary rocks from the present to a given time in the past.
KUMMEL, B. (1961). *History of the earth*. Freeman, San Francisco.
RONOV, A. B. (1964). Common tendencies in the chemical evolution of the earth's crust, ocean and atmosphere. *Geochemistry* **8**, 715–43. The data used in this book is adapted from that given in Kummel's book.

Total maximum thickness, in km of all deposits of ages zero to stated value in 10^6 yr: 5, 11; 11, 25; 19, 40; 28, 60; 32, 70; 48, 135; 60, 180; 70, 225; 76, 270; 82, 305; 90, 350; 102, 400; 112, 440; 124, 500; 136, 600; 144, 675 reaching the asymptotic value, estimated here, of 210 km. There were pronounced high rates of production now and in Jurassic and Devonian time, and corresponding low rates in Cretaceous, Carboniferous, and late Precambrian time. [118]

geophysics RUNCORN, S. K. (ed.) (1967). *International dictionary of geophysics*. Pergamon Press, Oxford.

geosyncline AUBOUIN, J. (1965). *Geosynclines*. Elsevier, Amsterdam.
A long depression filling with sediments and often subsequently strongly folded. Extremely variable objects but broadly having the spatial relationship: craton, mio-geosyncline, eu-geo-syncline. The craton, or so-called foreland, an older stable area, has relative movement trans-versely towards the troughs. The mio-geosyncline, adjacent to the craton, may have shales, graywackes (typically 10 km thick), basic and ultrabasic volcanic rocks, pulsatory activity and overall movement generally away from the craton, and a complex heating sequence. [119]

geothermal power [182, Fig. 15.13; Questions 97, 98]

geothermal areas [Chapter 15]

geyser Intermittent spouting spring; named after Geysir, Iceland, eruption period irregular, typically hours to a few days, spouting about 50 m. Dominant variables: water temperature and SVP cf. magma geyser description of Chatper 14 with pressure and DGP (*see also* DGP). [160, 163, 176]

Gilbert [87]

glacier PATERSON, W. S. B. (1969). *The physics of glaciers*. Pergamon Press, Oxford.
Stream of ice, usually deeper than 20 m, with flow typically $1–100\,\text{m yr}^{-1}$. Mean velocity $\sim g \sin\theta h^2/3v$ on a slope of angle θ, depth of flow h, kinematic viscosity $v \sim 10^{10}\,\text{m}^2\,\text{s}^{-1}$. [23, Fig. 4.1]

global development *See* structure; thermal history.

granite Acid igneous rock composed of quartz, feldspar, and usually mica; an abundant plutonic rock.

granitic In this book used in its broadest sense, all the rock-substances ranging from granite and rhyolite to grandiorite and rhyodacite, and also used rather loosely as more or less synonymous with acidic, and including andesite. [18, 32, 35, 95, 123, 146; Table 14.2]

grandiorite Acid rock, the most abundant of granitic rocks and of igneous intrusive rocks of the continental crust, containing more plagioclase feldspar than granite.

granulite Metamorphic rock of low water content produced by the most intense temperature and pressures of regional metamorphism. [123]

gravimeter An instrument for measuring the local acceleration of gravity. Normally the instrument is arranged to measure only differences from an arbitrary datum of the vertical component of gravity. Present-day field instruments can distinguish differences of 0·01 mgal, about 10^{-8} of the total field. [17]

gravitation constant G.

gravitational acceleration g.

gravitational anomaly *See* Bouger anomaly.

gravitational energy Role in thermal evolution [81]

gravitational force $= GmM/r^2$ between two particles of masses, m and M, distance r apart. *See also* G.

gravity CAPUTO, M. (1967). *The gravity field of the earth*. Academic Press, New York. [7, 17, 19, 149]

gravity formula (1967) $g = g_e(1 + K\sin^2\phi)/(1 - e^2\sin^2\phi)^{\frac{1}{2}}$, where g_e is the equatorial gravity, $K = 0·00193166$, $e^2 = 0·0066946$, and ϕ is the latitude.

graywackes Rather fine-grained sedimentary rocks of granitic composition usually deposited on the ocean floor by turbidity currents to thicknesses often exceeding 10 km. [98]

Greenland *See also* Ilimaussaq, Quvnertussup. [109]

ground swell [157]

h Vertical thickness or depth of given structure. Also length-scale.

H Depth of a column.

$\mathscr{H}$ Enthalpy change (at core–mantle boundary) $= 1700$ kJ kg^{-1}. [38]

Hawaii, volcanic system [149; 150, Fig. 14.5, 153; 155, Fig. 14.8; 156]

heat LEE, W. H. K. (ed.) (1965). *Terrestrial heat flow*. American Geophysical Union, Geophysical Monograph, No. 8, Washington, D.C.

heat and mass transfer [50; 130]

heat flux f. [30; 31, Fig. 5.2; 69; 72; 80, Fig. 9.2; 82, Table 9.2; 84, Table 9.3; 93, Fig. 10.4; 114; 134; 167, Fig. 15.3; 172, Fig. 15.7; 174; Fig. 15.8]

Hekla volcano, Iceland [159, Fig. 14.12]

Hele-Shaw cell The original study by H. S. J. HELE-SHAW (1898) in *Trans. Instn. nav. Archit.* **50**, 25 is described in the books by LAMB and SCHLICHTING (*see* fluid dynamics). The use of a Hele-Shaw cell for non-homogeneous fluids has been exploited notably by R. A. WOODING (for example (1960): *J. Fluid Mech.* **7**, 501–15) and the author (for example (1967): *J. Fluid Mech.* **27**, 609–23; (1967): *J. Fluid Mech.* **32**, 69–96). [130; 168]

hertz Hz $= s^{-1}$, SI unit of frequency, events per second.

HFU Heat flow unit $= 1\,\mu\text{cal cm}^{-2}\text{s}^{-1} = 41\cdot86\,\text{mW m}^{-2}$; a common unit for describing the net heat flux out of the earth.

Himalayas [33, Question 72]

hinterland In orogeny, the moving block which pushes the sediments towards the foreland.

historical geology SEYFERT, C. K. and SIRKIN, L. A. (1973). *Earth history and plate tectonics*. Harper and Row, New York.

homogeneous The same at every point. Thus a uniform medium is homogeneous and isotropic.

hot pools [177]

hot springs [166, 173]

Huygens [6]

hydraulic diameter *See* hydrodynamic resistance.

hydrodynamic resistance Due to relative motion of a more-or-less rigid body in a fluid. An unfortunate expression since the fluid need not be water. The forces arise from the effects of viscosity, pressure variations around bluff bodies, and the radiation of various waves such as those on a free surface or in a compressible medium. A voluminous subject because of its ubiquitous practical importance. For flow in a channel or pipe, including the case of a free surface, a sufficiently accurate empirical generalization in practice is as follows. Define the hydraulic diameter $d = 4$ (area of cross-section of flow)/(the length of wetted perimeter), and the Reynolds number $(Re) = Ud/v$, where U is the mean or discharge velocity and v the kinematic viscosity of the fluid. Then for a fluid of density ρ in a conduit under the action of a pressure gradient p' we find: $p'd/\frac{1}{2}\rho U^2 = \lambda$, where λ is solely a function of (Re) and a measure of the conduit roughness. For very rough conduits $\lambda \approx 0.01 + 32/(Re)$, the expression used in this book. Note that the hydrodynamic wall stress is about $p'd/4$. [50, 147, 161]

hydrosphere Units, 10^{17} kg; total 17 200: ocean 13 700 (80 per cent); pore water in sediments 3300 (18·8 per cent); ice 200 (1·2 per cent); rivers, lakes 0·3 (0·2 per cent); atmosphere 0·13 (0·08 per cent). Ocean-water recycle time 42 000 yr. Mass of dissolved matter into ocean $= 4.2 \times 10^{12}$ kg yr^{-1}. Entire dissolved mass in ocean replaced every 1.2×10^7 yr. Budget, unit 10^3 km^3 yr^{-1}: evaporation from ocean 360; from land 70; precipitation into ocean 330; land 100; run-off 30.

hydrostatic equilibrium [8; 19; 92; 148]

hydrothermal eruption *See* phreatic eruption.

hydrothermal systems ELDER, J. W. (1966). *Hydrothermal systems*. Bn169, New Zealand Department of Science and Industrial Research, Wellington.

  heat and mass transfer in the crust by the movement of water substance [Chapter 15]
  heat flow and temperature distribution [165, Figs 15.2–15.5]; with discharge [170, Figs 15.6, 15.7]
  surface zone [173]
  time scales [181]

hypsographic profile A relationship giving the area of the earth's solid surface above a given height, 100 per cent at -10.96 km (Mariana trench) to 0 per cent at 8·79 km (Everest), with mean level at -2.44 km. The mean land elevation is 0·84 km, mean sea-depth 3·8 km. There are two distinct most frequent levels: 0·27 km, -4.4 km. *See also* land, ocean.

Hz *See* hertz.

i Integer.

I Moment of inertia: principal values A, B, C.

ice Solid water, polycrystalline. Crystal structure of linked tetrahedra forming layers of puckered hexagonal rings. Several phases, all denser than water, are found above 2 kbar. Grain-size of new ice is typically 1–20 mm; old glacier ice up to 50 cm. Density at 0 °C, 0·917 g cm^{-3}; latent heat 334 kJ kg^{-1}; melting-point decreases 7·5 K kbar^{-1}. Semiconductor. Kinematic viscosity: single crystals 10^7 m^2 s^{-1}; new polycrystalline ice 10^{10} m^2 s^{-1}. P-wave velocity, 3·8 km s^{-1}. [26]

Ice Age Three major episodes currently identified: Cambrian 570 million years B.P.; Permo-carboniferous 280 million years B.P.; Pleistocene 3 million years B.P., until now. Pleistocene dates in 10^3 yr B.P.: Wurmian 10–70, peak at 18; Warthian 100–120; Rissian 175–210, Mindelian 370–600; Gunzian 750–1000. [20]

ice-sheet or cap Extensive ice deposits, notably in Antarctica and Greenland and smaller ones in Iceland, Spitsbergen, and the Arctic islands. $H^3 \sim 9vwl^2/g$: for height H in centre of sheet of total width $2l$; accumulation rate w, kinematic viscosity $v \sim 10^6$ m^2 s^{-1}. [20]

igneous rock Formed by heat, leading to partial or complete melting and subsequent crystallization. The simplest chemical classification of igneous rocks, if the chemical composition is represented as a mixture of oxides, is based on their equivalent silica content, which is usually the most abundant (*See also* crust): acidic, $>66\%$; intermediate, 52–66%; basic, 45–52%; ultrabasic $<45\%$. The more basic rocks have increasing proportions of ferromagnesian minerals, are darker, denser, more potentially magnetic and with higher seismic velocities.

ignimbrite A welded tuff normally formed by a volcanic ash flow of the *nuée ardente* type. Such flows or ash clouds may cover distances of about 100 km. [134, 163]

IGY International Geophysical Year 1957–9, a co-operative study of the earth by about 30 000 scientists at 2000 stations. Similar events were held in 1882–3, 1932–3.

Ilimaussaq intrusion, South Greenland [142, Fig. 13.12]

implosion [28]

inertia *See* moment of inertia.

instability *See* stability.

integer *i, n.*

interaction [50, 95]

interface [139]

interior flow [65, 70]

   temperature fluctuations [65, 76]

intrusion NEWALL, G. and RAST, N. (eds) Mechanism of igneous intrusion. *J. Geol.* (Special Issue No. 2.) [Chapter 13; Questions 79, 81]

ionic radii In ångströms (10^{-10} m): O^{2-}, 1·32; Si^{4+}, 0·39; Fe^{2+}, Fe^{3+}; 0·82; 0·67; Ca^{2+}, 1·06; Na^{+}, 0·98; K^{+}, 1·33; Mg^{2+}, 0·78.

Irish sea [111]

island arc [98]

isostasy The tendency of crustal masses to float in the mantle. [19–22; 118; Questions 19–23, 71]

isotropic At a given point, the same in all directions; cf. homogeneous.

Italy *See* Pozzuoli, Larderello, Tuscany, Vesuvius.

J *See* joule.

Japan *See* Aso-san, Kyushu, Oshima-san.

jelly model [109]

joint A crack across and along which there has been little or no displacement.

joule J, unit of work, energy and heat, $= N\,m = m^2\,kg\,s^{-2} = (1/4\cdot186)$cal, equal to work done by a force of 1 N moved 1 m.

k Permeability.

K Thermal conductivity in units $W\,m^{-2}\,K^{-1}$: sediments, 1·5; crust, both basaltic and granitic, 2; mantle 4. [30]

K *See* kelvin.

κ Thermal diffusivity $= K/\rho c$, nominally $10^{-6}\,m^2\,s^{-1}$.

Karipiti fumarole, New Zealand [177; 182, Fig. 15.13]

Karroo, Africa [151]

kelvin K, SI base unit of thermodynamic temperature. Defined to give the triple point of water substance as 273·16 K. This point is also defined at 0·01 °C. Note, for a temperature difference, 1 K = 1 °C. In general, temperatures in this book are in kelvins; this includes: absolute thermodynamic temperature, temperature differences, and compound units involving temperature such as thermal capacity, etc. The only exception is for actual temperatures in degrees Celsius (= Centigrade): example, 'basalt melts at 1100 °C.'

Kelvin, conduction model [85]

Kepler [7]

Kermadec *See* Tonga-Kermadec.

kg *See* kilogram.

Kilauea [150, Fig. 14.5; 154]; 1955 flank eruption [155, Fig. 14.8]. *See also* Hawaii.

kilogram kg, SI base unit of mass. Defined by an artifact – a particular lump of metal. Approximately the mass of 10^3 cm^3 of liquid water.

kimberlite Complex fragmented intrusive rock occurring in vertical funnels and dykes in the continental crust. Presumed to contain the deepest samples from the mantle. The kimberlites of southern Africa were erupted in Cretaceous time. [32; Question 30]

kinematic viscosity v. *See* viscosity.

Kyushu, Japan [30; 31, Fig. 5.1]. *See also* Aso-san.

l Length, arc length, width.

L Latent heat.

λ Wavelength. Radioactive decay constant, as in $y = y_0 \exp(-\lambda t)$.

laminar flow (of a fluid) Flowing smoothly, not turbulent.

land Total area 1.49×10^8 km^2, mean height 0.84 km. Areas of major land masses in 10^8 km^2 and mean height in km: Asia 0·44, 0·96; Africa 0·30, 0·75; North America 0·24, 0·72; South America 0·18, 0·59; Europe 0·10, 0·34; Australia 0·09, 0·34. *See also* hypsographic profile.

Laplace [41]

Larderello, Italy [179]

latent heat L. Energy required per unit mass for a specified phase change, about $300\,\mathrm{kJ\,kg^{-1}}$ for rock-substance.

lava pile [148]
   age [150]
   elevation [148]
   layering [153]
   mass [150]
   shape [150]
   plateau [148; Question 91]

layered intrusions [139, Figs 13.10–13.12; Question 81]

length, or width l.

length scale, of upper mantle [76]

light, speed of 2.998×10^8 m s^{-1}.

limiting velocity [57]

lithophilic elements [45]

lithostatic load The total weight of the material above the object of interest. [149, 150, 155, 161]

lithothermal system Heat and mass transfer by the movement of rock or magma. [Chapters 13, 14]

lithosphere A rather vague term referring to the outer, stiffer part of the earth, including the crust and part of the upper mantle. A useful notion which is unfortunately rarely defined with any precision.

m *See* metre.

m Mass, of secondary body; mass enclosed in sphere of radius r. As subscript m = mantle.

M Mass of earth, 6×10^{24} kg.

Maclaurin Flattening of a uniform rotating globe, $e = 15\omega^3/16\pi\rho \approx 1/230$ for the earth. Actual value 1/298 is smaller, thus since $e \propto 1/\rho$ for a non-homogeneous globe, we need ρ to increase with depth. [8]

magma A plastic mass or paste of solid and liquid silicate melts which occur within the earth, generally permeating surrounding rock, often containing gases, mostly H_2O, in solution, and crystals in suspension. [144–6, Table 14.2; 161]

magma geyser [157]

magma reservoir [131]

magnetic time-scale The times of reversal of the earth's magnetic field as shown in the normal and reversed direction of magnetization of lavas and oceanic sediments. Currently known to about 80 million years B.P. Taking the present as normal, reversals occurred at (in units of 10^6 yr B.P.): 0·69, 0·89, 0·95, 1·61, 1·63, 1·64, 1·79, 1·95, 1·98, 2·11, 2·13, 2·43, 2·80, 2·90, 2·94, 3·06, 3·32, 3·70, 3·92, 4·05.... [87]

magnetic stripes *See* ocean floor.

magnetism
JACOBS, J. A. (1963). *The earth's core and geomagnetism.* Pergamon Press, Oxford.
CHAPMAN, S. and BARTELS, J. (1962). *Geomagnetism.* Clarendon Press, Oxford.
IRVING, E. (1964). *Palaeomagnetism and its application to geological and geophysical problems.* John Wiley, New York and London.
RIKITAKE, T. (1966). *Electromagnetism and the earth's interior.* Elsevier, Amsterdam.

magnetite Fe_3O_4, the predominant magnetic mineral.

magnetohydrodynamics Study of the motion of a viscous, electrically conducting fluid, for example, in the core of the earth.

magnetometer Instrument for measuring the strength of a magnetic field or one of its components. Portable instrument precision is normally about 10^{-5} of the total field.

mantle Major solid outer part of the earth extending to depth 2900 km. Density range: $3·3$–$5·7$ g cm^{-3}. Mass $= 4·0 \times 10^{24}$ kg $= 0·67 \times$ mass of the earth. The mantle is a ubiquitous participant throughout this book.

mass m, M.

mass of earth M [7; Questions 4, 6]

masses and dimensions of parts of the earth [36, Table 5.1]

Mauna Loa *See* Hawaii.

mechanical equivalent of heat *See* calorie.

mechanical properties JAEGER, J. C. (1962). *Elasticity fracture and flow.* Methuen, London.
The mechanical properties of solid rock-substance are represented in a very idealized way in this book. Rock in lumps larger than a kilometre or so in scale are considered to have: elastic properties, for example with Young's modulus of order 1 Mbar, but incompressible except on a global scale; short-duration finite strength, for example with a shear strength somewhat less than 1 kbar; and viscosity of order 10^{20} P or so. In order to understand which of these properties will dominate a particular phenomenon it is important to realize that both the size of the system and the duration of the driving forces play a vital role. Many elementary books which refer only to the stress–strain relationships are quite misleading in this respect. Put at its simplest, elastic effects are transmitted at seismic velocity—the times involved are geologically negligible, whereas viscous processes, being rate dependent, take time to manifest themselves. Thus for example a pile of rock 10 km high, 200 km across, and of viscosity 10^{22} P will subside under its own weight by lateral viscous flow with a time-scale of order 10^8 yr. For times much less than this the shape of the pile will be dominated by its finite strength. The bigger the lump the shorter the time-interval—proportional to v/gh, where h is the length scale—before viscous processes dominate. On the global scale the geological effects of finite strength are negligible.

high-frequency properties [11]
low-frequency properties [17]
model of rock-substance [23]
melt(ing) [122]
basalt and granite [145, Fig. 14.2; 146, Table 14.2]
front [123]
partial [122; 126; 147]

proportion [151]
temperature [32]

metamorphic rock Obtained from pre-existing rock-substance by physical conditions—pressure, temperature, strain, wetness, etc.—differing in kind, intensity, or duration from those existing earlier. Metamorphism is not just a dramatic transformation, such as found in the contact metamorphism adjacent to a dyke, but is in effect a persistent and continuous process affecting the microstructure and fabric of all rock-substance all the time. Since it is difficult to find any rock-substance without some metamorphism the term is very broad. In its simplest usage the term is used in contrast to fresh sedimentary rock, formed perhaps directly by erosion or to fresh igneous rock. A good example of metamorphic rocks are schists.

metamorphism [122]; global [47]

meteorite Stone from space. Of 700+ observed falls and collection, about 85 per cent are stony, 15 per cent irons. Probably formed independently at the same time as the solar system. Ages 4.5×10^9 yr. Widely overrated as being representative of the rock-substance of the earth.

meteorological perspective [70]

metre m, SI base unit of length. The distance from the pole to the equator is very nearly 10^4 km. Defined by the wavelength of a radiation from atomic krypton-86 [6]

microseism Background motion of the solid earth, normally studied in the frequency band 10^{-2}–10^2 Hz, with vertical velocities generally much less than $1 \mu m s^{-1}$.

mid-ocean ridge A global system of swells on the ocean floor typically 2 km high and 1000 km broad. An inapt name: they are not particularly ridge-like nor exclusively centred in oceanic basins.

milligal *See* gravitational acceleration and gal.

mineral Naturally occurring crystalline component of solid rock-substance.

mixing [49]

moho Slang for Mohorovičič discontinuity (1909), named after its discoverer, where there is an abrupt increase with depth of *P*-wave velocity from about 7.6–8.6 km s^{-1}. Difficult to measure to better than 1 km accuracy in depth; 5 km in oceanic areas; 30 km in shield areas; up to 70 km beneath Himalayas. [19]

moment of inertia I, A, B, C; a measure of the inertial effects of a rotating body. For a single body rotating at angular velocity ω the angular momentum $I\omega$ is constant if no torque acts on the body. Its rotational energy is $\frac{1}{2}I\omega^2$. For a uniform sphere $I = \frac{2}{5}Ma^2$. For earth $C = 0.3308\, Ma^2$. Residuals; after Caputo: $(C-A)/Ma^2 = 1.99 \times 10^{-5}$, $(C-B)/Ma^2 = 1.27 \times 10^{-5}$; based on the hydrostatic term $C_{20} = 480.15 \times 10^{-6}$. [7, 91]
  orientation [92, Fig. 10.3; Question 12]

moon Mean distance from earth, 3.84×10^5 km; sidereal period, 27.3 day; radius, 1738 km; mass, 7.354×10^{22} kg = earth mass/81.28; $g = 1.62$ m s^{-2}; mean density 3.35 g cm^{-3}.

mountain chain [Chapter 12]. *See* orogenesis.

MPD The melting-point at a given depth.

μ Viscosity, nominal value for the upper mantle 3×10^{20} P.

mud pool, mud volcano [164, Fig. 15.1; 176]

'mushroom' [Chapter 13; 168]

n Integer, 3 here.

N *See* newton.

nappe A sheet-like body of rock typically of order 1 km thick, which has moved over the ambient surface, distances typically of order 10 km, usually with substantial structural deformation. Associated with vigorous orogenesis. Crudely analogous to a glacier of rock, but not to a land slide or slip. [26].

newton N, unit of force = m kg s^{-2} = 10^5 dyn, approximately the *weight* of a 100 g mass.

Newton, gravitational theory [6, 7]

New Zealand *See* Karipiti, Rotorua, Taupo, Tongariro, Wairakei, White Island.

normal area of heat flux [48]

v Kinematic viscosity $= \mu/\rho$, nominal value for the upper mantle $10^{16}\,\mathrm{m^2\,s^{-1}}$. [20; 80, Fig. 9.2; 82, Table 9.2]

(Nu) Nusselt number, dimensionless heat flux.

nuée ardente Most violent form of volcanic eruption ejecting a hot fluidized ash mass of great mobility.

Nusselt number (Nu) [52, 53, 58, 130, 171]

ocean Total area $3.62 \times 10^8\,\mathrm{km^2}$; volume $= 1.35 \times 10^9\,\mathrm{km^3}$; mean depth $= 3.8\,\mathrm{km}$. Salinity (per cent) 3.5 ± 0.1 of which Cl $= 1.94$, Na $= 1.08$, $SO_4 = 0.27$, Mg $= 0.13$.

Areas of major oceans in $10^8\,\mathrm{km^2}$: Pacific, 1.65; Atlantic, 0.82; Indian, 0.73; Arctic, 0.14; the rest 0.28. *See also* hypsographic profile.

ocean(ic) floor, spreading, ages of magnetic stripes [89, Fig. 10.2]
   hydrothermal system [172]
   ridge [108; Question 66]
   upper mantle [111]
   volcanic system [148]

ω Angular velocity.

orbital period T.

origin [41]

orogen A mountainous system; complex long prismatic part of the crust, showing major deformation of one or more stratigraphical units by folding, thrusting, and plutonism. [Chapter 12]

KENT, P. E., SATTERWAITE, G. E., and SPENCER, A. M. (eds) (1969). *Time and place in orogeny*. Geological Society, London.

An extremely interesting global model of orogenic systems, even though based on an arguably dubious use of studies of rotating and convecting fluid bodies, is found in DEARNLEY, R. (1966). Orogenic fold-belts and a hypothesis of earth evolution. *Physics Chem. Earth* **7**, 1–114. [Questions 67, 69, 72–8]
   cooking model [122, Fig. 12.4; 124, Table 12.3]
   overturning model [124; 125, Fig. 12.5]. *See also* centrifuge technique.
   time-scales [121]

Oshima-san volcano, Japan, cooling time [Question 96]

p Pressure difference.

P Pressure, total. Also power from radioactive decay, nominal initial value $P_0 = 10^{-19}\,\mathrm{W\,kg^{-1}}$.

P *See* poise.

parsec Astronomical unit of ditance $= 3.0857 \times 10^{13}\,\mathrm{km}$.

Pa *See* pascal.

(Pe) Peclet number.

Pacific [109]

pascal Unit of pressure $= 1\,\mathrm{N\,m^{-2}} = 10^{-5}\,\mathrm{bar}$.

Peclet number (Pe). [51; 100; 111]

peridotite An ultrabasic, coarse, dark, heavy intrusive rock, rich in olivine and other ferromagnesian minerals and with less than 10 per cent feldspar. Readily altered to serpentine. Considered by many to be similar in composition to the rock-substance of the upper mantle.

permeability Property of a porous solid to allow passage of fluid. In the relation due to Darcy, $q = -(k/\mu)\nabla p$, k is the permeability. A common unit of permeability is the darcy or millidarcy; 1 darcy $= 0.987 \times 10^{-12}\,\mathrm{m^2}$. To be carefully distinguished from porosity. [129; 148; 171]; of a microstructure [129]; [Questions 66, 96]

petrology WYLLIE, P. J. (1971). *The dynamic earth.* Wiley, New York.
   Petrology is the study of rock-substances as chemical sytems. In the solid state an archetypal rock is defined as a distinctive aggregate of minerals. Rocks are normally considered in three broad classes: sedimentary, metamorphic, and igneous. Modern petrology is mostly concerned with microscopic systems. It is a subject suited to classroom teaching so that its role in modern geology has become over-emphasized. Those parts of geological knowledge which lie between the microchemical and macrophysical are presently areas of almost total ignorance and likely to be fruitful areas of future study.

phase change [38; 157; 173]

phreatic eruption Explosive disruption of the upper few hundred metres of crust caused by a very rapid nearly adiabatic conversion of hot crustal water and rock to steam and rock. No warning! [176]

Picard [6]

pipe model of hydrothermal system [181]

planets (r = orbital semi-major axis, in 10^8 km, a = mean radius of body, in 10^3 km, ρ = mean density, in g cm^{-3}, g = surface acceleration, in m s^{-2}). See Table G.1.

TABLE G.1

	r	a	ρ	g
Mercury	0·579	2·42	5·5	3·80
Venus	1·08	6·11	5·2	8·70
Earth	1·50	6·37	5·52	9·80
Mars	2·28	3·38	3·9	3·70
Jupiter	7·78	70·0	1·4	23·00
Saturn	14·3	58·4	0·71	9·10
Uranus	28·7	23·5	1·32	9·70
Neptune	45·0	22·3	1·63	13·50
Pluto	59	6·0	6·46	5·20

planetary evolution [83; 84, Table 9.3]

plate tectonics A conceptual model of the kinematics of the rearrangement of the crust based on the notion of an outer thin, rigid shell, pieces of which are moving uniformly and independently except along lines of creation and destruction. Currently the *hoch Kultur* of geophysics. 'Plate' is a very unfortunate choice of word because it falsely implies that one considers the shell-like region as rigid, whereas it is extremely weak with shear strength less than 1 kbar and effective viscosity of order 10^{23} P. In reading the literature of this subject the term plate should be mentally translated to 'a part of the crust and upper mantle in which there is near the surface a region of nearly uniform horizontal flow'. [50]

plateau volcanism *See* lava plateau.

plume [67; 71; 111]

plutonic Refers to processes deep in the crust, and usually applied to igneous rocks.

plutons [120, 124]

poise Unit of viscosity, 1 P = 0·1 N s m^{-2} = 1 dyn s cm^{-2}.

Poisson's ratio [15]

polar wandering [Chapter 10]
   Model after GOLDREICH, P. and TOOMRE, A. (1969). Some remarks on polar wandering. *J. Geophys. Res.* **74**, 2555–67.

porosity e, proportion of void/unit volume or proportion of area of holes per unit cross-sectional area. [129, 153]

porous medium convection *See* convection, permeable.

Poseidonius [6]

potassium [39]

Pozzuoli, Italy, mud pool [164, Fig. 15.1]

Prandtl number $(Pr) = v/\kappa$, named after Ludwig Prandtl (*see also* fluid dynamics), one of the great pioneers of modern continuum mechanics. His work and its later developments, although mostly behind the scenes, permeate this book. The ballroom-dancing analogue [91] is a hypothetical use of a photographic technique exploited by him and his colleagues for studies of eddying motions. [51]

precession Motion of the axis of rotation of a body subjected to a torque. For the simple case of the torque Q at right-angles to the axis, the precession period is $2\pi I \omega/Q$, where I is the axial moment of inertia and ω the angular velocity of rotation.

pressure p, P.

proton Mass $= 1 \cdot 673 \times 10^{-27}$ kg.

ψ Stream function.

P-wave Dilatational seismic wave. [13]

pump [150; 154; 156, Fig. 14.9]

q Velocity, velocity amplitude, volumetric velocity

Q Sharpness of resonance parameter, a measure of the persistence parameter, a measure of a vibration. $2\pi/Q = \Delta E/E$, where E is the energy of the vibration and ΔE is the loss of energy per cycle, so that $E(t) = E(0)\exp(-2t/Q)$. Thus, $Q = f/\Delta f$, where Δf is the width of the special line of peak frequency f. Steel 10^4; quartz 2400; rubber 24; the earth, 10^2–10^4, purely radial mode $_0S_0 > 10^4$, others typically around 400. Also discharge in $m^3 s^{-1}$.

quartz Crystalline silica, SiO_2.

Quvnertussup sermia, West Greenland, small glacier [23, Fig. 4.1]

r Radial coordinate, radial distance.

R Radius of given object.

(Ra) Rayleigh number.

rad Radian, SI unit of angle, $180/\pi = 57 \cdot 3°$.

radiative thermal conductivity For a simple grey medium $K = 16n^2\sigma T^3/3\varepsilon$, where n is the refractive index, σ is Stephan's constant, $5 \cdot 67 \times 10^{-8}$ W m^{-2} K^{-4}, T is the absolute temperature, ε the extinction coefficient, that is, the intensity $\propto \exp(-\varepsilon r)$. A minor contribution within the earth, where probably $\varepsilon \gtrsim 10^5$ m^{-1}. Dominant effect in theory of origin of earth from collapsing cloud.

radiative transfer [34]

radioactive elements Half-life $T_{\frac{1}{2}}$, in 10^{10} yr units unless otherwise stated and decay energy kJ kg^{-1} yr^{-1}. Note the decay constant $\lambda = (\ln 2)/T_{\frac{1}{2}}$, where quantity $\propto \exp(-\lambda t)$.

Half-life of H^3, 12·26 yr; ^{14}C, 5568 yr; ^{40}K, 0·133 (0·88); ^{87}Rb, 5; ^{115}In, 6×10^4; ^{138}La, 7; ^{147}Sm, $1 \cdot 25 \times 10^6$; ^{178}Lu, 2·4; ^{187}Re, 5; ^{232}Th, 1·39 (0·84); ^{235}U, 0·071 (18); ^{238}U, 0·45 (3·0).

radioactive heating [34; 39, Table 5.2; 81; 84]

radius a, c, r, R.

radius of earth a; [5; Question 3]

Ramberg, Hans *See* centrifuge technique, structural geology.

random function ε [93, 134, 154]

Rayleigh number (Ra); $(Ra)_m$ for a porous medium.

The early studies of the onset of weak convection are:

LAPWOOD, E. R. (1948). Convection of a fluid in a porous medium. *Proc. Camb. phil. Soc. math. phys.* **44**, 508–21.

RAYLEIGH, LORD (1916). On convection currents in a horizontal layer of fluid when the higher temperature is on the underside. *Phil. Mag.* **32**, 529–46.

The topic is treated at length in Chapter 2 of CHANDRASEKHAR, S. (1961). *Hydrodynamic and hydromagnetic stability.* Clarendon Press, Oxford.

The Rayleigh number for onset of convection between horizontal planes depends on the boundary conditions on the two planes: both rigid, 1708; one rigid, 1101; both free (zero stress), 658. This situation is directly not the slightest interest in global geology, but is valuable in leading to an understanding of the role of molecular processes in the thermal sublayer. [53; 58; 60; 95; 101–6; 113; 121; 141]

permeable medium [130, 171]. Critical [58, 71, 130]

(Re) Reynolds number.

recharge of water [169; 170, Fig. 15.6]

reservoir [29, 131, 145, 153, 154, 166, 179]

resistance flow R In a pipe-like system defined by $p = qR$, where p is the pressure drop driving a volume flow per unit time, q through the system. For flow through ground of permeability k into a vent-like structure of depth h with fluid able to enter over the bulk of its depth $R = v/\xi kh$, where ξ is a geometrical factor of the order of unity. For viscous flow up a fissure of depth h, horizontal extent l, and gap-width a, $R = 12vh/la^3$. For viscous flow up a vent with impermeable walls $R = 8vh/\pi a^4$. [153; 175]

resistance, internal For flow from a reservoir into a vent of diameter d, the geometrical resistance factor $\xi \approx 2\pi/\ln(2h/d)$. Note that permeability is estimated, in this book, from $k = e^3 \Delta^2/48$, where e is the porosity and Δ the typical channel width of the passages in the permeable medium.

resonance [12]

resonance parameter Q.

Reynolds number (*Re*). [50]

Reynolds, O The foundation of the modern statistical approach to studies of turbulence was (1894): On the dynamical theory of incompressible fluids and the determination of the criterion. *Phil. R. Soc. Trans.* **A186**, 123. The stress components, such as $\rho\langle uv \rangle$ are known as the 'Reynolds stresses'. *See also* bladder model.

rheology [23]. *See also* bladder model.

ρ Density. When otherwise unspecified, earth's mean density = $5.517\,\mathrm{g\,cm}^{-3}$.

rhyolite An igneous rock, the volcanic equivalent of granite. Special forms are the glass, obsidian, and pumice. Commonly occurs in the continental crust, but small quantities are found among oceanic basalts. [134]

rift [109]

rock A naturally occurring consolidated piece of the solid earth, which has some distinct chemical or physical heterogeneity.

rock-substance For the want of a suitable term, I use *rock-substance* to refer to rock without implying that it is solid, liquid, or gaseous. This book treats the variety of rock-substance in an extremely idealized manner. Only six varieties are recognized: (1) rock-substance in general, specified solely by density and viscosity; (2) unconsolidated sediments, of granitic composition and solid density $2.4\,\mathrm{g\,cm}^{-3}$; (3) rock-substance of granitic composition of solid density $2.7\,\mathrm{g\,cm}^{-3}$ and extreme viscosity; (4) basaltic rock-substance of solid density $3.0\,\mathrm{g\,cm}^{-3}$ and moderate viscosity; (5) ultrabasic rock-substance of solid density $3.3\,\mathrm{g\,cm}^{-3}$; (6) a group of unknown dense, extremely basic rock-substances of solid densities exceeding $4\,\mathrm{g\,cm}^{-3}$ constituting the bulk of the mantle and core. [122, 128, 144]. *See also* equation of state; mechanical properties.

Roll-over time-scale [83, Fig. 9.3, 8.4, Table 9.3]

rotation MUNK, W. H. and MACDONALD, G. J. F. (1960). *The rotation of the earth.* Cambridge University Press. [46, 90]. *See also* time.

Rotorua, New Zealand, intrusive complex [135, Fig. 13.5]

s *See* second.

S Used as a general elastic modulus, $S \sim 1$ Mbar. Also solute concentration. Also thermal sublayer parameter.

salinity The mass proportion of a solute in a solution. For example, the salinity of sea-water and animal body fluids is about 0·03. [139]

salt dome Diapiric evaporite structures of horizontal scale a kilometre or more penetrating through overlying sediments by flow of the solid salts, in the simplest case due to their relative buoyancy. [131]

satellite, period of revolution about central body [7]

scale A measure of a quantity. Used in two senses. (1) *Specific*: an actual measure which unequivocally specifies the quantity; for example, the size of a sphere could be given by any one of radius, surface area, or volume. (2) *Order of magnitude*: a measure to indicate the typical magnitude of the quantity; for example, a day is about 10^5 s and journeys in a large city are on the 10 km scale. Any system has its own intrinsic scales. For example the time-scale to heat a body of length-scale h by conduction is h^2/κ, the time-scale of a pendulum is $(h/g)^{\frac{1}{2}}$. If we express the measures of a system in units based on its intrinsic scales, then all examples of that class of systems are directly comparable. *See also* dimensional analysis.

Scotland, Caledonian orogeny [120, Table 12.2; 121, Fig. 12.3]

second s, SI base unit of time. Defined by the frequency of a radiation from atomic caesium. Approximately the period of a 25-cm-long pendulum.

sedimentary rock Accumulated deposits of chemical, biochemical and mainly fragmental origin subsequently compacted, cemented, and lithified. Typical examples: evaporites, chalk, sandstone. [119; 120, Table 12.2]

sedimentation An excellent introduction to global sedimentary processes is: GARRELS, R. M. and MACKENZIE, F. T. (1971). *Evolution of sedimentary rocks.* Norton, New York. *See also* erosion.

sedimentation rate Deep ocean-floor sediments accumulated at about 1 cm per 10^3 yr = 1 km per 10^8 yr. *See also* erosion.

segregation
  of substance of earth [43]
  of crust [95]

seismic wave nomenclature P, compressional wave (after primary, first arrival); S, torsional wave (after secondary, later arrival); K, compressional wave in core; I, compressional wave in inner core; J, torsional wave in inner core; c, mantle–core–mantle reflection. Note that at an interface a wave may change its mode, P to S, or S to P.
Examples:
$PKIKP$ purely compressional wave through inner core
ScP torsional wave reflected at the core boundary as a compressional wave
$PKKP$ compressional wave reflected inside core on the other side
$PKPPKP$ compressional wave through core reflected back from the crust through the core

seismic velocity Of compressional wave in rock, typical values in km s^{-1} at 10 kbar: granite, 6·45; granodiorite, 6·56; quartz–diorite, 6·71; gabbro, 7·24; dunite, 8·15; eclogite, 7·87; graywacke, 6·20; slate, 6·22; amphibolite, 7·35.

seismology BULLEN, K. E. (1963). *Introduction to the theory of seismology.* Cambridge University Press, London. [11, 19, 149]

seismometer An instrument for measuring short-term displacements, from milliseconds to years, of the solid earth. The commoner instruments are velocity or acceleration transducers for detecting surface vibrations in the range 100–0·01 Hz. Velocities of the order of 10^{-9} m s^{-1} and above are readily measurable. The commonest such instrument is the geophone, a small robust transducer, typically working in the range 10–100 Hz, widely used in exploration seismology. Other seismometers, especially for very long periods, measure relative displacement directly. [26]

self-propulsion *See* crust, migration.

shadow zone [14]

shear stress τ

sheet [82]

SI System of measurement based on the six basic units: metre, kilogram, second, kelvin, ampere, candela. Originally formulated at the General Conference on Weights and Measures, 1960.

σ Poisson's ratio.

silicate Natural silicon–oxygen compounds of various metals, arranged as tetrahedra of four oxygens and one silicon atom in single tetrahedron or double tetrahedra or in chains, sheets, rings, and elaborate framework structures. The earth is a collection of silicates with a minor amount of other material.

sill An igneous intrusion horizontally extensive; when of lenticular form and domed, a laccolith; when centrally thinned, a lopolith. [152]. *See also* centrifuge technique.

slab [50]

    gravity formula [18]

slip plane [26]

Snell [6]

solar system Mass, 2×10^{30} kg; radius, 6×10^9 km; semi-minor axis of equivalent oblate spheroid, 3×10^7 km; mean density, 5×10^{-10} g cm^{-3}.

solute concentration S, Σ.

specific heat c Increase of internal energy per unit mass per K, for most rock-substance about $1 \text{ kJ kg}^{-1} \text{ K}^{-1}$.

spectra of turbulent convection [67–8]

spring *See* hot springs.

stability [55, 59, 124, 137, 139]

state *See* equation of state.

steam The vapour or gaseous phase of water-substance. Treated in the models of Chapters 14 and 15 as a perfect gas. Density $= \xi p/T$, where p is the pressure in bar, T is the absolute temperature and $\xi = 0.22$, 0.24, 0.31 (g cm^{-3}) K bar^{-1} at $100\,°C$, $200\,°C$, $300\,°C$ respectively. Otherwise the steam-tables are used to obtain ξ or density directly.

steaming ground [177; Question 95]

Stokesian flow A flow dominated by viscous forces and in which the effects of inertia are negligible. The limiting case of very slow flows. [50, 176]

strain A measure of the local distortion of a material. A field of strain can be specified by the relative displacements of nearby points. A pair of nearby points can be represented by a line vector, as can its displacement. Thus 9 components are required to specify the strain field at a point; for example $u(i,j)$, the proportional displacement in the j-direction of points along the i-direction. Strain is a three-dimensional tensor with zero dimensions, normally defined as $2e(i,j) = u(i,j) + u(j,i)$ representing the pure strain, together with a local rigid-body rotation $2\omega(i,j) = u(i,j) - u(j,i)$. In elasticity theory ω plays a minor role.

stream function ψ. A measure of the volume of fluid being transported per unit time. In steady flow, fluid particles follow streamlines of constant ψ.

strength Stress at which specified failure of a solid occurs. Typically of order 0.1 kbar for hard materials. [25; 144; 149–51; 152; 154; Questions 26, 86, 87]

stress A measure of the local force system in a material, obtained from the force acting on one side of a unit area imbedded in the material. Since the measured force will in general be a function of the orientation of the test area, a complete specification of the stress in three-dimensional space requires 9 components $p(i,j)$, where $i,j = 1, 2, 3$, corresponding, for example, to the three perpendicular directions x, y, z, $p(i,j)$ for the i component of the force measured on a test area with its normal pointing in the j-direction. Stress is a three-dimensional tensor with the dimensions of pressure. In an elastic material stress is a linear function of strain. In a viscous material stress is a linear function of *rate of strain*.

structural geology
RAMBERG, H. (1967). *Gravity deformation and the earth's crust.* Academic Press, New York.
RAMSEY, J. G. (1967). *Folding and fracturing of rocks.* McGraw-Hill, New York.

DE SITTER, L. U. (1964). *Structural geology*. McGraw-Hill, London.

structure, global development [Chapter 6]

subduction zone A crustal sump. *See* Benioff zone.

sublayer [58; Chapter 8]

sun Energy source, conversion of matter to energy mainly from 4 protons → 1 helium nucleus -0.0277 a.m.u., releasing energy c^2 per unit mass, where $c = $ velocity of light, that is, 6×10^4 J per kg of protons. Total radiation at effective temperature 5785 K over surface area of 6.09×10^{18} m^2 is 3.9×10^{20} MW. Flux at earth's mean distance 1.39 kW m$^{-2} = 2$ cal cm^{-2} min^{-1}. Total input to earth 2×10^{11} MW. Stellar magnitude -26.6; spectral class, G. Mass 1.99×10^{30} kg; radius 6.96×10^5 km.

surface Distribution on earth, in units of 10^6 km^2: total, 510; land, 149 (or 29·2 per cent); ocean, 361 (or 70·8 per cent); continental shield, 105; young folded belts, 42; volcanic islands in deep oceanic and sub-oceanic region, 2; shelves and continental slopes, 93; deep oceanic region 268 (*see also* hypsographic profile).

surface zone
   hydrothermal system [145, Table 14.1; 173]
   lithothermal systems [145, Table 14.1]

SVP Saturated vapour pressure, a function of temperature; the relation between the pressure and temperature of a vapour in thermodynamic equilibrium with its liquid. In this book used solely for water. [175]

S-wave Torsional seismic wave. [13]

Système Internationale *See* SI.

t Elapsed time.

R Temperature. Also period, orbital period.

$\mathcal{T}$ *See* temperature.

Tane-rore [67]

τ Decay time, usually for $y = y_0 \exp(-t/\tau)$. Also shear stress.

taxonomic Classification of objects by means of systematic relations between the objects themselves, rather than by their mode of occurrence with other objects. Especially used for classification of organisms. This book has a taxonomic approach.

Taupo area, New Zealand [2, Figs 1.1, 134; 135, Fig. 13.5; 165, Fig. 15.2, 167, Fig. 15.3; 171]

tectonic Refers to large-scale structures and the mechanical processes involved in forming and modifying them.

temperature In general, T. Where the temperature field is unsteady or varies in space or both and it is necessary to distinguish between the actual temperature and various averages the temperature is given the symbol $\mathcal{T}$. This notation is found only in Chapter 8. Nominal values: temperature in general, for simple rough estimates, initial mean global temperature 3000 °C; mantle below the sublayer 2500 °C; scale of temperature anomalies for meso-scale processes 1000 K. The figures of 3000 °C and 2500 °C are most likely too large by perhaps as much as 500 °C. *See also* kelvin.

temperature anomaly θ.

temperature difference across sublayer ϑ.

temperature distribution in hydrothermal system [167–9]

temperature of interior [31; 32, Fig. 5.3; 72; 77, Fig. 8.14; Questions 30, 61]
   profile [72]

temperature of surface of earth Mean value 287·4 K (14·2 °C). In this book normally taken in very rounded figures as 300 K or 0 °C. Variation: zonal averages at latitude 0 (10) 90° N or S, in °C; 26·2, 26·0, 24·1, 19·4, 13·0, 5·8, $-2·3$, $-12·2$, $-22·1$, $-27·9$. The southern hemisphere is considerably colder than the northern; the above figures are an average of the temperatures in both zones. The average range is 54 K, or about ± 10 per cent of the mean.

thermal area [31]

thermal capacity of earth 5×10^{27} J K^{-1}.

thermal conduction CARSLAW, H. S. and JAEGAR, J. C. (1969). *Conduction of heat in solids.* Clarendon Press, Oxford. [30, 33]
  cooling of sphere [33; 34, Fig. 5.4]
  models [33, 84]

thermal conductivity K.

thermal diffusivity κ.

thermal history [Chapter 9; Questions 31, 32, 38, 39, 50–4, 56]

thermal resistance A measure of the difficulty of transfer of heat, by analogy with the flow of electric current as stated in Ohm's law. The resistance of a slab of thickness l, cross-sectional area A, and thermal conductivity K is $R = 1/KA$. Thus the total power transferred $Q = \Delta T/R$, where ΔT is the temperature difference across the slab. For a set of slabs: in series, $R =$ sum of the individual resistances; in parallel, $1/R =$ sum of the individual reciprocal resistances. It is important to realize that the concept of thermal resistance is not restricted to a conductive system. Note that the thermal resistance will be increased by the presence of internal heating, and reduced by internal convection. [98, 100, 126]

thermal turbulence [Chapter 8; Table 8.1]
  mechanistic models [72–6]
  statistical properties [65–70]
  visualization [161, Figs 8.1–8.3]

thermosolutal convection [139]
TURNER, J. S. (1973). *Buoyancy effects in fluids.* Cambridge University Press.

θ Angle, general, normally in radians. Also temperature, variation; $\theta' =$ r.m.s. temperature variation.

ϑ Temperature across the thermal sublayer.

thickness, vertical h.

thorium [39]

Tibesti, North Africa [29]

tidal interaction Alteration of the earth's rotation rate owing to the action of tidal forces on the mass of the earth (including the ocean and atmosphere). [40, 47]

tide of the solid earth MELCHIOR, P. (1966). *The earth tides.* Pergamon Press, Oxford.
  For an ideal earth the change in gravity $\Delta g/g = (m'/m)(a/r)^3$, where m is the mass of the earth, m' of the other body, a the radius of the earth, and r the orbital radius of the other body. For the moon $\Delta g/g = 5.7 \times 10^{-8}$, and the sun 2.6×10^{-8}.

time t. One sidereal day $= 86\,164$ s; 1 sidereal year $= 3.156 \times 10^7$ s, 1 tropical year $= 365.2422$ days. In Devonian time, from coral-ring counts: 1 yr ≈ 400 days, 13 lunar months each of 30.5 days. Thus the relative change in rotation rate is $\Delta \omega / \omega \approx 0.025/10^8$ yr.
WELLS, J. W. (1963). Coral growth and geochronology. *Nature, Lond.* **197**, 948–50.

time, geological, beginning of [42]

time-scale, geological (after Holmes) Beginning of period (in units of 10^6 yr). Quaternary, 3; Tertiary, 65; Cretaceous, 136; Jurassic, 190; Triassic, 225; Permian, 280; Carboniferous, 345; Devonian, 395; Silurian, 430; Ordovician, 500; Cambrian, 570; Archean, 5000. The individual epochs: Pliocene, 7; Miocene, 26; Oligocene, 38, Eocene, 54; Paleocene, 65.
HARLAND, W. B., SMITH, A. G., and WILCOCK, B. (eds) (1964). *The phanerozoic time-scale* (a symposium). Geological Society, London.
  This time-sequence is based on the different types of fossil which preserve a record of the biosphere. Because the time-scale of dispersion of organisms in the biosphere is geologically very rapid this time-sequence is more-or-less worldwide. We cannot, however, assume that other geological effects will occur nearly simultaneously worldwide. On the contrary, macrogeological processes have a time-scale of order 100×10^8 yr so that in general their effects will at any given time occur approximately at random over the surface. *See also* time-scales.

vibrations *See* free vibrations.

vigour [81]

viscosity (dynamic) A measure of the rate of strain in an imperfectly elastic material subject to a distortional stress; in the case of a simple shear $\tau = \mu(\partial u/\partial y)$, where $u(y)$ is the velocity at y and μ is the viscosity. For systems changing in time, the kinematic viscosity $v = \mu/\rho$ with dimensions $m^2 s^{-1}$ is more appropriate.

Kinematic viscosities ($m^2 s^{-1}$) of common substances are: water, 10^{-6}, porridge, $0\cdot1$; asphalt at 20 °C, 10^2. For rock-substances: solid mantle as a whole, 10^{16}; liquid granite, 10^2; liquid basalt, $0\cdot1$.

Liquid rock at 1400 °C (P): hornblende–granite, 2×10^6; hornblende–mica–andesite, 2×10^4; andesite, 500; diorite, 80.

Variation with temperature: Mihara basalt 1951, °C (P): 1038 °C, 230; 1083 °C, 71×10^3; 1108 °C, 18×10^3; 1125 °C, $5\cdot6 \times 10^3$. [17; 129; 145; Question 27]
    crustal load model [20]
    effect of temperature and equivalent silica content [49]
    kinematic, of rock substance [27, Fig. 4.3]
    temperature variation [27, 81]
    variation, effect on structure [137]

visco-elastic model [23]

viscous stress [33, 52, 104]

volatiles [41, 143]

volcano
BULLARD, F. M. (1962). *Volcanoes*. University of Texas Press, Austin.
MACDONALD, G. A. (1974). *Volcanoes*. Prentice-Hall, New Jersey.
STEARNS, H. T. (1966). *Geology of the State of Hawaii*. Pacific Books, Palo Alto.
COATS, R. R., HAY, R. L., and ANDERSON, C. A. (eds) (1968). *Studies in volcanology*, Memoir 116, Geological Society, Colorado, America.
    eruptions Typical intervals between eruptions: Hawaii, 11 yr with 130-yr modulation. Etna, 8 yr with 55-yr modulation; Vesuvius, 20 yr, extremely variable in range 5–40 yr. [100, 107–8, 111, 114, 126, Chapter 14; Questions 83–7, 90, 96]

vorticity A term used in fluid mechanics; twice the effective local angular velocity of the fluid. [51, 62]

w Vertical velocity

W *See* watt.

Wairakei, New Zealand, thermal area [206; 2, Fig. 1.1; 29; Chapter 15]

wake [51]

water FAXEN, O. H. (1953). *Thermodynamic tables in the metric system for water and steam*. Nordisk Rotogravyr, Stockholm.
    water-substance Physical properties of liquid phase (*see also* ice, steam): density at 4 °C, $1\,g\,cm^{-3}$; 100 °C, $0\cdot9584\,g\,cm^{-3}$. Viscosity: at 0 °C, 17·8 mP, at 100 °C, 2·8 mP. Surface tension, $0\cdot072\,N\,m^{-2}$; self-diffusion, $2\cdot6 \times 10^9\,m^2\,s^{-1}$; heat of vaporization, $2\cdot26\,J\,kg^{-1}$; velocity of sound, $1\cdot5\,km\,s^{-1}$; thermal conductivity, $0\cdot59\,W\,m^{-1}\,K^{-1}$. Incompressibility, $0\cdot022\,M$bar. *See also* hydrosphere. [144–5; 157; 166; 173]

watt $W = J\,s^{-1}$, unit of power = rate of working or use of energy.

wavelength λ.

wave velocity c, α, β.

weakness of rock *See* strength.

White Island, New Zealand [165]

WHP Well-head pressure. For natural discharge systems, 1 bar.

x Horizontal or circumferential coordinate.

ξ A ratio of specified quantities for convenience in local working only. Also amplitude of surface undulation.

y Transverse, horizontal, or circumferential coordinate.

year *See* time.

Young's modulus [15]

z Vertical or radial coordinate, normally downward from surface.

ζ Dimensionless viscous drag coefficient, $\frac{2}{9}$ for a sphere.